Greenhouse Gases: Impact and Management

Greenhouse Gases: Impact and Management

Isabelle Mullins

MURPHY & MOORE

www.murphy-moorepublishing.com

Published by Murphy & Moore Publishing,
1 Rockefeller Plaza,
New York City, NY 10020, USA
www.murphy-moorepublishing.com

Greenhouse Gases: Impact and Management
Isabelle Mullins

International Standard Book Number: 978-1-63987-271-8 (Hardback)

Cataloging-in-Publication Data

Greenhouse gases : impact and management / Isabelle Mullins.
   p. cm.
Includes bibliographical references and index.
ISBN 978-1-63987-271-8
1. Greenhouse gases. 2. Greenhouse gas mitigation. 3. Environmental management.
4. Environmental impact analysis. I. Mullins, Isabelle.
TD885.5.G73 G74 2022

363.738 74--dc23

TABLE OF CONTENTS

It is with great pleasure that I present this book. It has been carefully written after numerous discussions with my peers and other practitioners of the field. I would like to take this opportunity to thank my family and friends who have been extremely supporting at every step in my life.

The gases that absorb and emit energy of gravitational and electromagnetic radiation within the thermal infrared range are known as greenhouse gases. These gases play an important role in the maintenance of the temperature of the Earth. They are the main cause of the greenhouse effect. Some of the greenhouse gases that are present in the Earth's atmosphere are carbon dioxide, nitrous oxide, water vapour and ozone. The impact of these individual gases on the total greenhouse effect depends upon their atmospheric lifetime, global warming potential and proportion of direct effects at a given moment. Burning of fossil fuels like coal, agricultural activities and use of chlorofluorocarbons are the sources of emission of these gases. Various natural processes and negative emissions of these gases are some of the ways to control their release in the atmosphere. This book unfolds the innovative aspects related to this area which will be crucial for the holistic understanding of the subject matter. Different approaches, evaluations, methodologies and studies related to this field have been included in this book. It will provide comprehensive knowledge to the readers.

The chapters below are organized to facilitate a comprehensive understanding of the subject:

Chapter – Introduction

Gases which absorb and emit infrared radiations are called greenhouse gases. These gases are responsible for causing the greenhouse effect, which results in global warming and other related environmental issues. This chapter sheds light on the aspects of greenhouse gases to provide a brief understanding of the subject.

Chapter – Common Greenhouse Gases

There are many types of greenhouse gases namely carbon dioxide, carbon monoxide, ozone, methane, tetrafluoromethane, nitrous oxide, chlorodifluoromethane, fluoroform, trifluoromethyl sulfur pentafluoride, chlorofluorocarbons, sulfur hexafluoride, etc. The topics elaborated in this chapter will help in gaining a better perspective of the different types of greenhouse gases.

Chapter – Negative Impacts of Greenhouse Gas

Greenhouse gases are responsible for affecting the environment adversely in many ways. Greenhouse effect, runaway greenhouse effect, enhanced greenhouse effect, global warming, radiative forcing, etc. are some of these effects. This chapter has been carefully written to provide an easy understanding of the various effects of greenhouse gases.

Chapter – Mitigation and Management

Practices of sustainable energy, sustainable transport, greenhouse gas removal, mobile emission reduction credit, carbon dioxide removal, fossil fuel phase-out, etc. are some of the mitigation and management methods used for greenhouse gas mitigation. This chapter closely examines the greenhouse gas mitigation and management practices to provide an easy understanding of the subject.

Chapter – Carbon Neutrality

Carbon neutrality refers to the process of nullifying carbon emission with carbon removal to achieve net zero carbon dioxide emissions. It includes concepts of carbon accounting, carbon offset, carbon sequestration, carbon capture and storage, carbon footprint, etc. This chapter covers all the related concepts of carbon neutrality for a thorough understanding of the subject.

Isabelle Mullins

Introduction

Gases which absorb and emit infrared radiations are called greenhouse gases. These gases are responsible for causing the greenhouse effect, which results in global warming and other related environmental issues. This chapter sheds light on the aspects of greenhouse gases to provide a brief understanding of the subject.

GREENHOUSE GASES

Greenhouse gases (GHGs) is the name given to a number of gases present in the earth's atmosphere which reduce the loss of heat into space and therefore contribute to global temperatures through the greenhouse effect. These gases are essential to maintaining the temperature of the Earth and without them the planet would be so cold as to be uninhabitable.

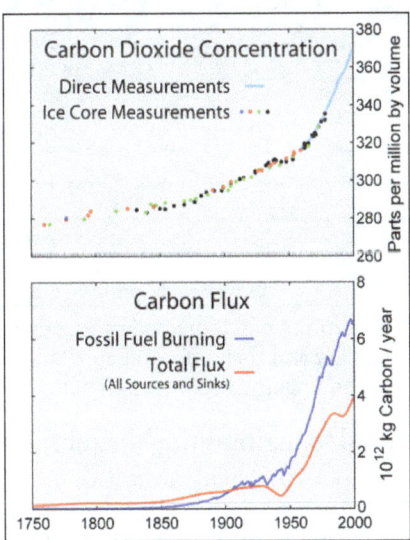

Top: Increasing atmospheric CO_2 levels as measured in the atmosphere and ice cores. Bottom: The amount of net carbon increase in the atmosphere, compared to carbon emissions from burning fossil fuel.

The most important greenhouse gas is water vapor which creates clouds. The vast bulk of this is produced by the natural process of evaporation of water from the sea. An excess of greenhouse gases can raise the temperature of a planet to lethal levels, as on

Venus where the 90 bar partial pressure of carbon dioxide (CO_2) contributes to a surface temperature of about 467 °C (872 °F). Carbon dioxide is produced by many natural and industrial processes, which currently result in CO_2 levels of 380 ppmv in the atmosphere. The reason for such a low level of carbon dioxide is that CO_2 is quickly taken up by plants through photosynthesis and converted into carbohydrates.

Based on ice-core samples and records current levels of CO_2 are approximately 100 ppmv higher than during immediately pre-industrial times, when direct human influence was negligible.

The Greenhouse Effect

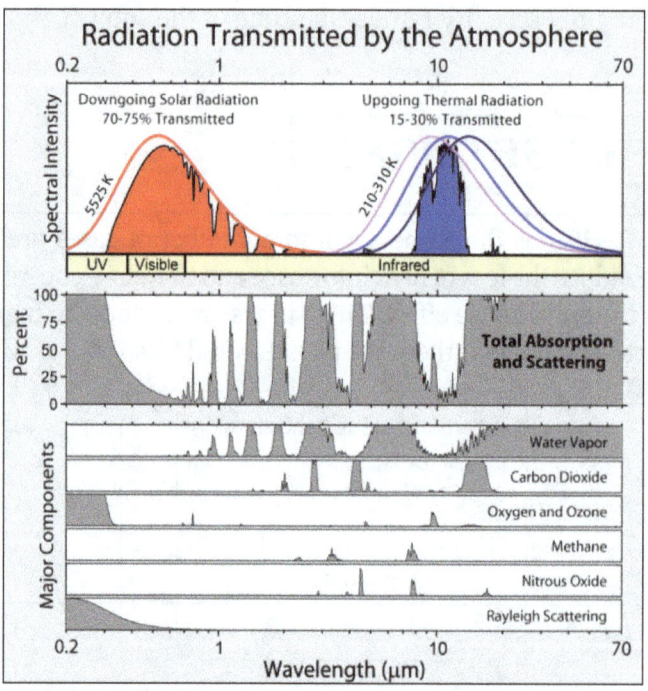

Pattern of absorption bands created by greenhouse gases
in the atmosphere and their effect on both solar radiation
and upgoing thermal radiation.

"Greenhouse gases" are essential to maintaining the temperature of the Earth—without them the planet would be so cold as to be uninhabitable.

When sunlight reaches the surface of the Earth, some of it is absorbed and warms the surface. Because the Earth's surface is much cooler than the sun, it radiates energy at much longer wavelengths than the sun does, peaking in the infrared at about 10 μm. The atmosphere absorbs these longer wavelengths more effectively than it does the shorter wavelengths from the sun. The absorption of this longwave radiant energy warms the atmosphere; the atmosphere is also warmed by transfer of sensible and latent heat from the surface.

Greenhouse gases also emit longwave radiation both upward to space and downward to the surface. The downward part of this longwave radiation emitted by the atmosphere is the "greenhouse effect." The term is a misnomer though, as this process is not the mechanism that warms greenhouses.

On earth, the most abundant greenhouse gases are, in order of relative abundance:

- Water vapor.
- Carbon dioxide.
- Methane.
- Nitrous oxide.
- Ozone.
- CFCs.

The most important greenhouse gases are:

- Water vapor, which causes about 36–70 percent of the greenhouse effect on Earth. (Note that clouds typically affect climate differently from other forms of atmospheric water.)
- Carbon dioxide, which causes 9–26 percent.
- Methane, which causes 4–9 percent.
- Ozone, which causes 3–7 percent.

This is a combination of the strength of the greenhouse effect of the gas and its abundance. For example, methane is a much stronger greenhouse gas than CO_2, but present in much smaller concentrations.

It is not possible to state that a certain gas causes a certain percentage of the greenhouse effect, because the influences of the various gases are not additive. (The higher ends of the ranges quoted are for the gas alone; the lower ends, for the gas counting overlaps.) Other greenhouse gases include, but are not limited to, nitrous oxide, sulfur hexafluoride, hydrofluorocarbons, perfluorocarbons and chlorofluorocarbons. A significant greenhouse gas not yet addressed by the IPCC (or the Kyoto Protocol) is nitrogen trifluoride.

The major atmospheric constituents (nitrogen, N_2 and oxygen, O_2) are not greenhouse gases. This is because homonuclear diatomic molecules such as N_2 and O_2 neither absorb nor emit infrared radiation, as there is no net change in the dipole moment of these molecules when they vibrate. Molecular vibrations occur at energies that are of the same magnitude as the energy of the photons on infrared light. Heteronuclear diatomics such as CO or HCl absorb IR; however, these molecules are short-lived in the atmosphere owing to their reactivity and solubility. As a consequence they do not contribute significantly to the greenhouse effect.

Late nineteenth-century scientists experimentally discovered that N_2 and O_2 did not absorb infrared radiation (called, at that time, "dark radiation") and that CO_2 and many other gases did absorb such radiation. It was recognized in the early twentieth century that the known major greenhouse gases in the atmosphere caused the earth's temperature to be higher than it would have been without the greenhouse gases.

Natural and Anthropogenic

Most greenhouse gases have both natural and anthropogenic sources. During the pre-industrial holocene, concentrations of these gases were roughly constant. Since the industrial revolution, concentrations of all the long-lived greenhouse gases have increased due to human actions.

Gas	Preindustrial Level	Current Level	Increase since 1750	Radiative forcing (W/m²)
Carbon dioxide	280 ppm	384ppm	104 ppm	1.46
Methane	700 ppb	1,745 ppb	1,045 ppb	0.48
Nitrous oxide	270 ppb	314 ppb	44 ppb	0.15
CFC-12	0	533 ppt	533 ppt	0.17

Ice cores provide evidence for variation in greenhouse gas concentrations over the past 800,000 years. Both CO_2 and CH_4 vary between glacial and interglacial phases, and concentrations of these gases correlate strongly with temperature. Before the ice core record, direct measurements do not exist. Various proxies and modeling suggests large variations; 500 millions years ago CO_2 levels were likely 10 times higher than now. Indeed higher CO_2 concentrations are thought to have prevailed throughout most of the Phanerozoic eon, with concentrations four to six times current concentrations during the Mesozoic era, and ten to fifteen times current concentrations during the early Palaeozoic era until the middle of the Devonian period, about 400 million years ago. The spread of land plants is thought to have reduced CO_2 concentrations during the late Devonian, and plant activities as both sources and sinks of CO_2 have since been important in providing stabilizing feedbacks. Earlier still, a 200-million year period of intermittent, widespread glaciation extending close to the equator (Snowball Earth) appears to have been ended suddenly, about 550 million years ago, by a colossal volcanic outgassing which raised the CO_2 concentration of the atmosphere abruptly to 12 percent, about 350 times modern levels, causing extreme greenhouse conditions and carbonate deposition as limestone at the rate of about 1mm per day. This episode marked the close of the Precambrian eon, and was succeeded by the generally warmer conditions of the Phanerozoic, during which multicellular animal and plant life evolved. No volcanic carbon dioxide emission of comparable scale has occurred since. In the modern era, emissions to the atmosphere from volcanoes are only about 1 percent of emissions from human sources.

Anthropogenic Greenhouse Gases

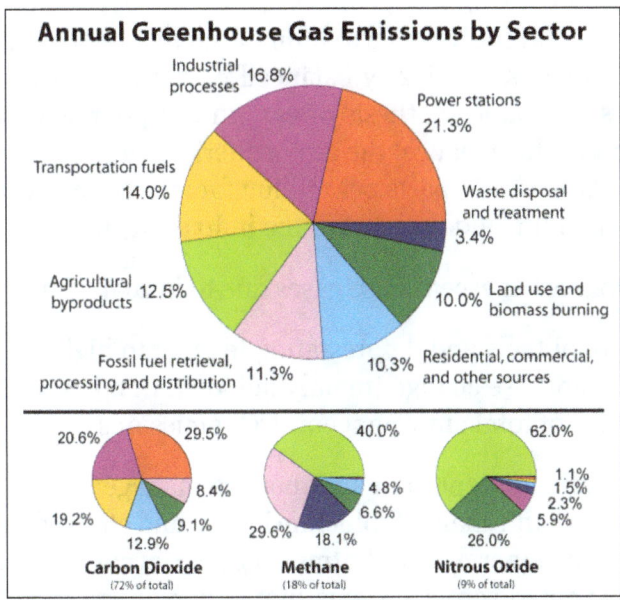

Global anthropogenic greenhouse gas emissions broken down into 8 different sectors for the year 2000.

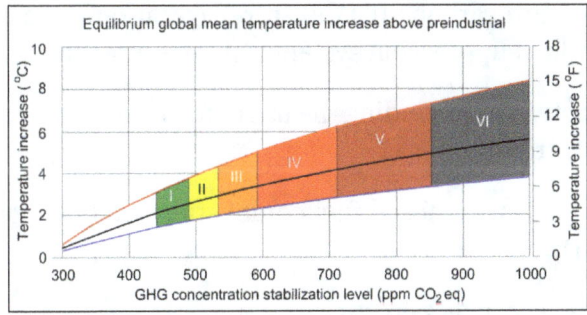

The projected temperature increase for a range of greenhouse gas stabilization scenarios (the colored bands). The black line in middle of the shaded area indicates 'best estimates'; the red and the blue lines the likely limits. From the work of IPCC AR4, 2007.

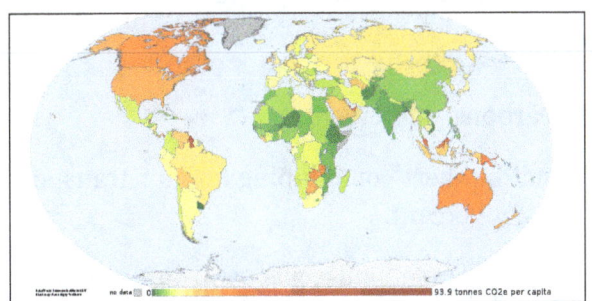

Per capita anthropogenic greenhouse gas emissions by country for the year 2000 including land-use change.

Since about 1750, human activity has increased the concentration of carbon dioxide and of some other important greenhouse gases. Natural sources of carbon dioxide are more than 20 times greater than sources due to human activity, but over periods longer than a few years natural sources are closely balanced by natural sinks such as weathering of continental rocks and photosynthesis of carbon compounds by plants and marine plankton. As a result of this balance, the atmospheric concentration of carbon dioxide remained between 260 and 280 parts per million for the 10,000 years between the end of the last glacial maximum and the start of the industrial era.

Some of the main sources of greenhouse gases due to human activity include:

- Burning of fossil fuels and deforestation leading to higher carbon dioxide concentrations. Land use change (mainly deforestation in the tropics) account for up to one-third of total anthropogenic CO_2 emissions.

- Livestock enteric fermentation and manure management, paddy rice farming, land use and wetland changes, pipeline losses, and covered vented landfill emissions leading to higher methane atmospheric concentrations. Many of the newer style fully vented septic systems that enhance and target the fermentation process also are sources of atmospheric methane.

- Use of chlorofluorocarbons (CFCs) in refrigeration systems, and use of CFCs and halons in fire suppression systems and manufacturing processes.

- Agricultural activities, including the use of fertilizers, that lead to higher nitrous oxide concentrations.

The seven sources of CO_2 from fossil fuel combustion are (with percentage contributions for 2000–2004):

- Solid fuels (e.g. coal): 35 percent.

- Liquid fuels (e.g. gasoline): 36 percent.

- Gaseous fuels (e.g. natural gas): 20 percent.

- Flaring gas industrially and at wells: <1 percent.

- Cement production: 3 percent.

- Non-fuel hydrocarbons: <1 percent.

- The "international bunkers" of shipping and air transport not included in national inventories: 4 percent.

The U.S. EPA ranks the major greenhouse gas contributing end-user sectors in the following order: industrial, transportation, residential, commercial and agricultural. Major sources of an individual's GHG include home heating and cooling, electricity consumption, and transportation. Corresponding conservation measures are improving home building insulation, compact fluorescent lamps and choosing energy-efficient vehicles.

Carbon dioxide, methane, nitrous oxide and three groups of fluorinated gases (sulfur hexafluoride, HFCs, and PFCs) are the major greenhouse gases and the subject of the Kyoto Protocol, which came into force in 2005.

Although CFCs are greenhouse gases, they are regulated by the Montreal Protocol, which was motivated by CFCs' contribution to ozone depletion rather than by their contribution to global warming. Note that ozone depletion has only a minor role in greenhouse warming though the two processes often are confused in the media.

Role of Water Vapor

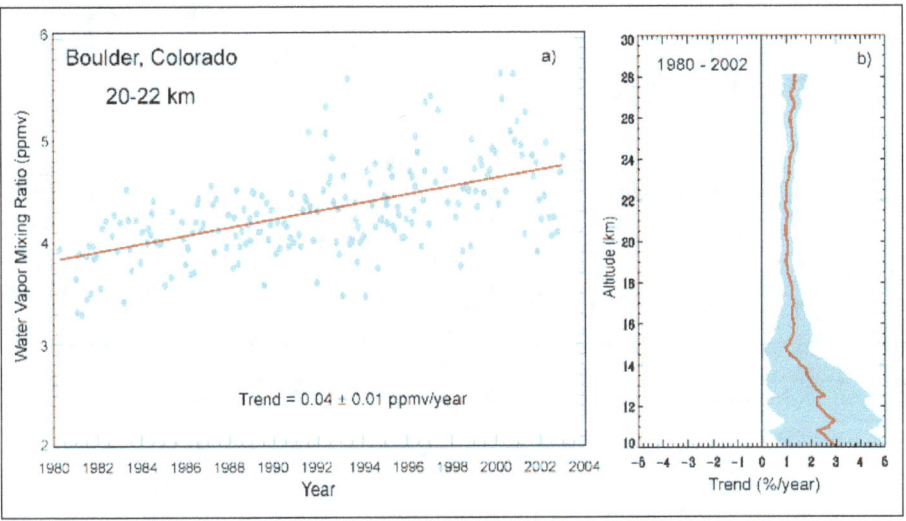

Increasing water vapor at Boulder, Colorado.

Water vapor is a naturally occurring greenhouse gas and accounts for the largest percentage of the greenhouse effect, between 36 percent and 66 percent. Water vapor concentrations fluctuate regionally, but human activity does not directly affect water vapor concentrations except at local scales (for example, near irrigated fields).

The Clausius-Clapeyron relation establishes that warmer air can hold more water vapor per unit volume. Current state-of-the-art climate models predict that increasing water vapor concentrations in warmer air will amplify the greenhouse effect created by anthropogenic greenhouse gases while maintaining nearly constant relative humidity. Thus water vapor acts as a positive feedback to the forcing provided by greenhouse gases such as CO_2.

Greenhouse Gas Emissions

Measurements from Antarctic ice cores show that just before industrial emissions started, atmospheric CO_2 levels were about 280 parts per million by volume (ppm; the units μL/L are occasionally used and are identical to parts per million by volume). From the same ice cores it appears that CO_2 concentrations stayed between 260 and 280 ppm during the preceding 10,000 years. However, because of the way air is trapped in ice and the time period represented in each ice sample analyzed, these figures are long term averages not annual levels. Studies using evidence from stomata of fossilized leaves suggest greater variability, with CO_2 levels above 300 ppm during the period 7,000–10,000 years ago, though others have argued that these findings more likely reflect calibration/contamination problems rather than actual CO_2 variability.

Since the beginning of the Industrial Revolution, the concentrations of many of the greenhouse gases have increased. The concentration of CO_2 has increased by about 100 ppm (i.e., from 280 ppm to 380 ppm). The first 50 ppm increase took place in about 200 years, from the start of the Industrial Revolution to around 1973; the next 50 ppm increase took place in about 33 years, from 1973 to 2006. Many observations are available online in a variety of Atmospheric Chemistry Observational Databases. The greenhouse gases with the largest radiative forcing are:

Relevant to radiative forcing				
Gas	Current Amount (1998) by volume	Increase over pre-industrial (1750)	Percentage increase	Radiative forcing (W/m²)
Carbon dioxide	365 ppm {383 ppm(2007.01)}	87 ppm {105 ppm(2007.01)}	31% {37.77%(2007.01)}	1.46 {~1.532 (2007.01)}
Methane	1,745 ppb	1,045 ppb	150%	0.48
Nitrous oxide	314 ppb	44 ppb	16%	0.15

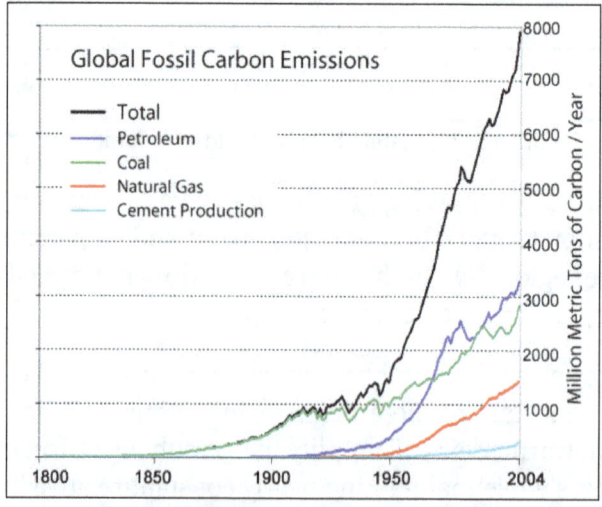

Global anthropogenic Carbon emissions.

Relevant to both radiative forcing and ozone depletion; all of the following have no natural sources and hence zero amounts pre-industrial		
Gas	Current (1998) Amount by volume	Radiative forcing (W/m²)
CFC-11	268 ppt	0.07

CFC-12	533 ppt	0.17
CFC-113	84 ppt	0.03
Carbon tetrachloride	102 ppt	0.01
HCFC-22	69 ppt	0.03

Long-term Trend

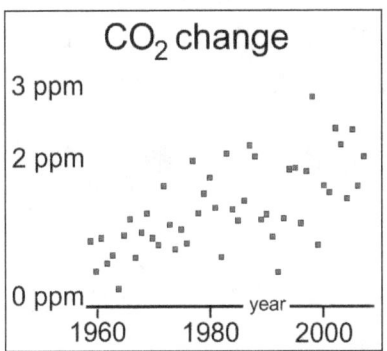

Year-to-year increase of atmospheric CO_2.

Atmospheric carbon dioxide concentration is increasing at an increasing rate. In the 1960s, the average annual increase was only 37 percent of what it was in 2000 through 2007.

Removal from the Atmosphere and Global Warming Potential

Aside from water vapor, which has a residence time of days, it is believed that most greenhouse gases take many years to leave the atmosphere.

Greenhouse gases can be removed from the atmosphere by various processes:

- As a consequence of a physical change (condensation and precipitation remove water vapor from the atmosphere).

- As a consequence of chemical reactions within the atmosphere. This is the case for methane. It is oxidized by reaction with naturally occurring hydroxyl radical, OH• and degraded to CO_2 and water vapor at the end of a chain of reactions (the contribution of the CO_2 from the oxidation of methane is not included in the methane Global warming potential). This also includes solution and solid phase chemistry occurring in atmospheric aerosols.

- As a consequence of a physical interchange at the interface between the atmosphere and the other compartments of the planet. An example is the mixing of atmospheric gases into the oceans at the boundary layer.

- As a consequence of a chemical change at the interface between the atmosphere and the other compartments of the planet. This is the case for CO_2, which is reduced by photosynthesis of plants, and which, after dissolving in the oceans, reacts to form carbonic acid and bicarbonate and carbonate ions.

- As a consequence of a photochemical change. Halocarbons are dissociated by UV light releasing Cl• and F• as free radicals in the stratosphere with harmful effects on ozone (halocarbons are generally too stable to disappear by chemical reaction in the atmosphere).

- As a consequence of dissociative ionization caused by high energy cosmic rays or lightning discharges, which break molecular bonds. For example, lightning forms N anions from N_2 which then react with O_2 to form NO_2.

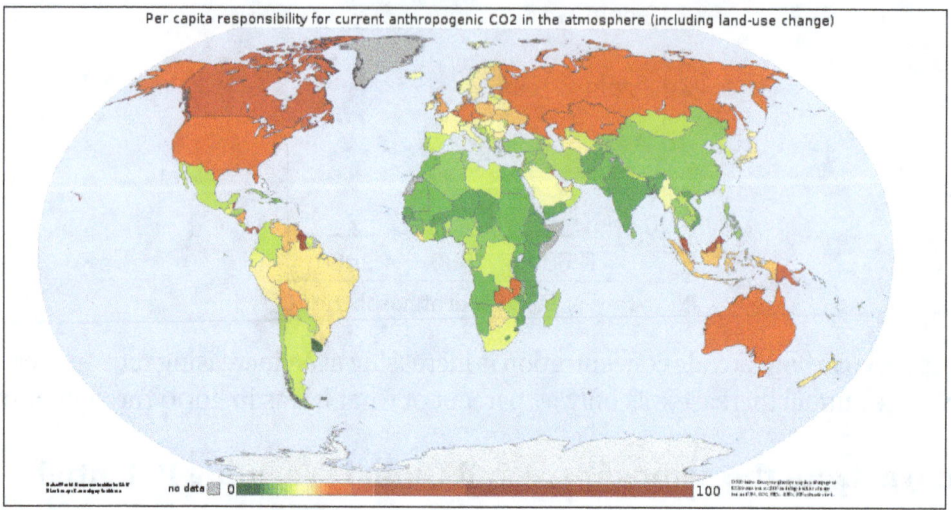

Per capita responsibility for current anthropogenic atmospheric CO_2.

Atmospheric Lifetime

Jacob defines the lifetime τ of an atmospheric species X in a one-box model as the average time that a molecule of X remains in the box. Mathematically τ can be defined as the ratio of the mass m (in kg) of X in the box to its removal rate, which is the sum of the flow of X out of the box (F_{out}), chemical loss of $X(L)$, and deposition of $X(D)$ (all in kg/sec): $\tau = \dfrac{m}{F_{out} + L + D}$

The atmospheric lifetime of a species therefore measures the time required to restore equilibrium following an increase in its concentration in the atmosphere. Individual atoms or molecules may be lost or deposited to sinks such as the soil, the oceans and other waters, or vegetation and other biological systems, reducing the excess to background concentrations. The average time taken to achieve this is the mean lifetime. The atmospheric lifetime of CO_2 is often incorrectly stated to be only a few years because that is the average time for any CO_2 molecule to stay in the atmosphere before being removed by mixing into the ocean, photosynthesis, or other processes. However, this ignores the balancing fluxes of CO_2 into the atmosphere from the other reservoirs. It is the net concentration changes of the various greenhouse gases by all sources and sinks that determines atmospheric lifetime, not just the removal processes.

Examples of the atmospheric lifetime and GWP for several greenhouse gases include:

- CO_2 has a variable atmospheric lifetime, and cannot be specified precisely. Recent work indicates that recovery from a large input of atmospheric CO_2 from burning fossil fuels will result in an effective lifetime of tens of thousands of years. Carbon dioxide is defined to have a GWP of 1 over all time periods.

- Methane has an atmospheric lifetime of 12 ± 3 years and a GWP of 62 over 20 years, 23 over 100 years and 7 over 500 years. The decrease in GWP associated with longer times is associated with the fact that the methane is degraded to water and CO_2 by chemical reactions in the atmosphere.

- Nitrous oxide has an atmospheric lifetime of 120 years and a GWP of 296 over 100 years.

- CFC-12 has an atmospheric lifetime of 100 years and a GWP of 10600 over 100 years.

- HCFC-22 has an atmospheric lifetime of 12.1 years and a GWP of 1700 over 100 years.

- Tetrafluoromethane has an atmospheric lifetime of 50,000 years and a GWP of 5700 over 100 years.

- Sulfur hexafluoride has an atmospheric lifetime of 3,200 years and a GWP of 22000 over 100 years.

The use of CFC-12 (except some essential uses) has been phased out due to its ozone depleting properties. The phasing-out of less active HCFC-compounds will be completed in 2030.

Airborne Fraction

Airborne fraction (AF) is the proportion of an emission (e.g. CO_2) remaining in the atmosphere after a specified time. Canadell define the annual AF as the ratio of the atmospheric CO_2 increase in a given year to that year's total emissions, and calculate that of the average 9.1 PgC y^{-1} of total anthropogenic emissions from 2000 to 2006, the AF was 0.45. For CO_2 the AF over the last 50 years has been increasing at 0.25±0.21%/year.

Global Warming Potential

The global warming potential (GWP) depends on both the efficiency of the molecule as a greenhouse gas and its atmospheric lifetime. GWP is measured relative to the same mass of CO_2 and evaluated for a specific timescale. Thus, if a molecule has a high GWP on a short time scale (say 20 years) but has only a short lifetime, it will have a large GWP on a 20 year scale but a small one on a 100 year scale. Conversely, if a molecule has a longer atmospheric lifetime than CO_2 its GWP will increase with time.

ADVANTAGES OF GREENHOUSE GASES

It is common knowledge that greenhouse gases cause global warming via the greenhouse effect. But people do not realize that these greenhouse gases are also responsible for sustaining life on Earth.

Greenhouse Gases as Barrier-filter

The greenhouse gases causing the greenhouse effect work as a barrier-filter for the atmosphere. The sun radiates solar energy on the earth and greenhouse gases make sure that 45% of these harmful solar radiations do not reach the surface of the planet by bouncing back most of the damaging UV radiation back into space.

Greenhouse Gases Maintain Ideal Temperature

Greenhouse gases help maintain an ideal habitable temperature on the earth surface. The reason Earth is habitable is due to the fact that the temperature level on Earth is maintained at an ideal level. This makes it possible for humans, plants, animals and other organisms to thrive. It is thanks to the greenhouse gases that the Earth is warm enough to achieve this ideal temperature. Without them, the Earth's surface would have ended up being a lot cooler than its current temperature.

Greenhouse Gases and Ozone

Ozone is one of the most crucial greenhouse gases. It is also the only gas that can prevent harmful ultraviolet (UV) rays from entering the Earth's atmosphere. Without the Ozone layer, UV rays could easily reach the surface and cause widespread damage, including various skin diseases.

Greenhouse Gases Maintain Water Level

It is thanks to the greenhouse effect that the Earth is able to maintain the water levels that exist on the surface. The planet has not melted completely due to the moderate temperature that exists on it. The polar ice caps also remain restricted to just the two polar regions of the Earth.

Agriculture and Greenhouse Gases

The greenhouse effect has enabled human agriculture practices to overcome the boundation of climate. Greenhouse gases allow farmers to grow seasonal crops whenever they want to. Greenhouse gases with the variations in temperatures synthetically create whichever season the farmer chooses. Help in heating water.

Hence, the greenhouse effect has helped maintain the perfect balance between the amount of absorbed energy and the amount of reflected energy in the atmosphere. However, it is due to the sudden increase in the use of fossil fuel since the industrial revolution that there has been an alarming rise in the number of greenhouse gases in the atmosphere of the Earth like carbon dioxide by 40%. This has led to the increase in the amount of energy absorbed and trapped in the atmosphere. This is also the reason for the anomaly termed as 'global warming' to come into being.

CARBON FOOTPRINT

Carbon footprint is the amount of carbon dioxide (CO_2) emissions associated with all the activities of a person or other entity (e.g., building, corporation, country, etc). It includes direct emissions, such as those that result from fossil-fuel combustion in manufacturing, heating, and transportation, as well as emissions required to produce the electricity associated with goods and services consumed. In addition, the carbon footprint concept also often includes the emissions of other greenhouse gases, such as methane, nitrous oxide, or chlorofluorocarbons (CFCs).

The carbon footprint concept is related to and grew out of the older idea of ecological footprint, a concept invented in the early 1990s by Canadian ecologist William Rees and Swiss-born regional planner Mathis Wackernagel at the University of British Columbia. An ecological footprint is the total area of land required to sustain an activity or population. It includes environmental impacts, such as water use and the amount of land used for food production. In contrast, a carbon footprint is usually expressed as a measure of weight, as in tons of CO_2 or CO_2 equivalent per year.

Carbon Footprint Calculation

Carbon footprints are different from a country's reported per capita emissions (for example, those reported under the United Nations Framework Convention on Climate Change). Rather than the greenhouse gas emissions associated with production, carbon footprints focus on the greenhouse gas emissions associated with consumption. They include the emissions associated with goods that are imported into a country but are produced elsewhere and generally take into account emissions associated with international transport and shipping, which is not accounted for in standard national inventories. As a result, a country's carbon footprint can increase even as carbon emissions within its borders decrease.

The per capita carbon footprint is highest in the United States. According to the Carbon Dioxide Information Analysis Center and the United Nations Development

Programme, in 2004 the average resident of the United States had a per capita carbon footprint of 20.6 metric tons (22.7 short tons) of CO_2 equivalent, some five to seven times the global average. Averages vary greatly around the world, with higher footprints generally found in residents of developed countries. For example, that same year France had a per capita carbon footprint of 6.0 metric tons (6.6 short tons), whereas Brazil and Tanzania had carbon footprints of 1.8 metric tons (about 2 short tons) and 0.1 metric ton (0.1 short ton) of CO_2 equivalent, respectively.

In developed countries, transportation and household energy use make up the largest component of an individual's carbon footprint. For example, approximately 40 percent of total emissions in the United States during the first decade of the 21st century were from those sources. Such emissions are included as part of an individual's "primary" carbon footprint, representing the emissions over which an individual has direct control. The remainder of an individual's carbon footprint is called the "secondary" carbon footprint, representing carbon emissions associated with the consumption of goods and services. The secondary footprint includes carbon emissions emitted by food production. It can be used to account for diets that contain higher proportions of meat, which requires a greater amount of energy and nutrients to produce than vegetables and grains, and foods that have been transported long distances. The manufacturing and transportation of consumer goods are additional contributors to the secondary carbon footprint. For example, the carbon footprint of a bottle of water includes the CO_2 or CO_2 equivalent emitted during the manufacture of the bottle itself plus the amount emitted during the transportation of the bottle to the consumer.

A variety of different tools exist for calculating the carbon footprints for individuals, businesses, and other organizations. Commonly used methodologies for calculating organizational carbon footprints include the Greenhouse Gas Protocol, from the World Resources Institute and the World Business Council for Sustainable Development, and ISO 14064, a standard developed by the International Organization for Standardization dealing specifically with greenhouse gas emissions. Several organizations, such as the U.S. Environmental Protection Agency, the Nature Conservancy, and British Petroleum, created carbon calculators on the Internet for individuals. Such calculators allow people to compare their own estimated carbon footprints with the national and world averages.

Carbon Footprint Reduction

Individuals and corporations can take a number of steps to reduce their carbon footprints and thus contribute to global climate mitigation. They can purchase carbon offsets (broadly stated, an investment in a carbon-reducing activity or technology) to compensate for part or all of their carbon footprint. If they purchase enough to offset their carbon footprint, they become effectively carbon neutral.

Carbon footprints can be reduced through improving energy efficiency and changing lifestyles and purchasing habits. Switching one's energy and transportation use can have an impact on primary carbon footprints. For example, using public transportation, such as buses and trains, reduces an individual's carbon footprint when compared with driving. Individuals and corporations can reduce their respective carbon footprints by installing energy-efficient lighting, adding insulation in buildings, or using renewable energy sources to generate the electricity they require. For example, electricity generation from wind power produces no direct carbon emissions. Additional lifestyle choices that can lower an individual's secondary carbon footprint include reducing one's consumption of meat and switching one's purchasing habits to products that require fewer carbon emissions to produce and transport.

Common Greenhouse Gases

2

There are many types of greenhouse gases namely carbon dioxide, carbon monoxide, ozone, methane, tetrafluoromethane, nitrous oxide, chlorodifluoromethane, fluoroform, trifluoromethyl sulfur pentafluoride, chlorofluorocarbons, sulfur hexafluoride, etc. The topics elaborated in this chapter will help in gaining a better perspective of the different types of greenhouse gases.

Carbon dioxide is a chemical compound composed of one carbon and two oxygen atoms. It is often referred to by its formula CO_2.

It is present in the Earth's atmosphere at a low concentration and acts as a greenhouse gas. In its solid state, it is called dry ice. It is a major component of the carbon cycle.

Atmospheric carbon dioxide derives from multiple natural sources including volcanic outgassing, the combustion of organic matter, and the respiration processes of living aerobic organisms; man-made sources of carbon dioxide come mainly from the burning of various fossil fuels for power generation and transport use.

It is also produced by various microorganisms from fermentation and cellular respiration. Plants convert carbon dioxide to oxygen during a process called photosynthesis, using both the carbon and the oxygen to construct carbohydrates.

Properties

Carbon dioxide is a simple covalent molecule that most people have heard about, as it is often in the news linked to global warming. Carbon dioxide has the formula CO_2 and at the centre of this linear molecule is a carbon atom joined by two pairs of double-bonds to the oxygen atoms, i.e O=C=O. At room temperature carbon dioxide is a colourless gas which has a slightly sweet smell. Carbon dioxide is a linear molecule with a bond angle of 180°. This gas was 'discovered' by Scottish scientist (physicist and chemist) Joseph Black.

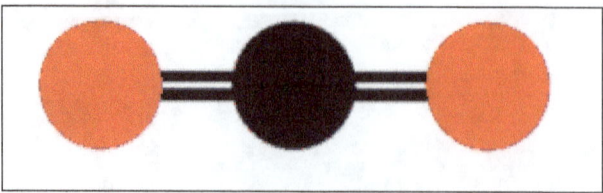

Carbon dioxide.

Dry Ice

Solid carbon dioxide is known by chemists as cardice and by everyone else as dry ice. It is an unusual solid as it sublimes (turning from solid to gas without going through the liquid state). Solid carbon dioxide has weak forces (van der Waals) between molecules which holds them together. This is why it has a very low melting point 217 K (-56 °C) at 5.2 atmospheres. Carbon dioxide is non-polar as its dipoles cancel, and this contrasts with the polar molecule sulfur dioxide (SO_2).

Fire extinguisherCarbon dioxide fire extinguishers contain highly pressurized carbon dioxide which is a non-flammable inert gas, which acts as a smothering material. They work by displacing the air (oxygen) from the area surrounding the fire. Once this part of the fire triangle has been removed the fire is extinguished. This type of fire extinguisher has the added advantage of also cooling the fuel. Carbon dioxide extinguishers are suitable for class B fires (flammable liquids and gases). These extinguishers have the advantage of not leaving a residue after they have been used (discharged). They are suitable for electrical fires in offices containing computers, televisions and photocopiers.

In the unlikely event that you are faced with a magnesium fire it is not advisable to use a carbon dioxide fire extinguisher because it actually reacts with magnesium.

$$2 \, Mg \, (s) \, + \, CO_2 \, (g) \rightarrow 2 \, MgO \, (s) \, + \, 2C \, (s)$$

Carbon dioxide will burn on reaction with magnesium and the result is the formation of magnesium oxide and carbon as soot. There are some superb films of the reaction online using blocks of dry ice (solid carbon dioxide). Carbon dioxide is a solid below -78°C. This is why dry ice should not be handled as it causes burns by freezing the skin.

Reactions

Carbon dioxide is an acidic oxide (a typical property of the majority of non-metal oxides) and reacts with sodium hydroxide to form a salt and water. Next time you are in the laboratory check the top of the sodium hydroxide bottles. If you see a white crust around the stoppers inform the technicians who need to supply fresh solutions. This simple chemistry has been exploited in craft such as submarines and space shuttles where the build up of carbon dioxide from respiration needs tackling.

$$CO_2 \, (g) \, + \, 2 \, NaOH \, (aq) \rightarrow Na_2CO_3 \, (aq) \, + \, H_2O \, (l)$$

The thermal decomposition of lithium carbonate and the following Group 2 carbonates leads to the production of carbon dioxide gas and the metal oxide. Interestingly, beryllium carbonate is unstable under standard conditions (298 K, 100 kPa) and the subsequent Group 1 carbonates are thermally stable.

$$Li_2CO_3 \, (s) \rightarrow Li_2O \, (s) \, + \, CO_2 \, (g) \quad (>1310°C)$$

$$MgCO_3(s) \rightarrow MgO(s) + CO_2(g) \quad (>350°C)$$

$$CaCO_3(s) \rightarrow CaO(s) + CO_2(g) \quad (>832°C)$$

$$SrCO_3(s) \rightarrow SrO(s) + CO_2(g) \quad (>1340°C)$$

$$BaCO_3(s) \rightarrow BaO(s) + CO_2(g) \quad (>1450°C)$$

The trend is that on descending Group 2 their carbonates become more difficult to decompose thermally as reflected by the increasing temperatures. These differences can be explained in terms of charge-density differences for the cations (M^{2+}).

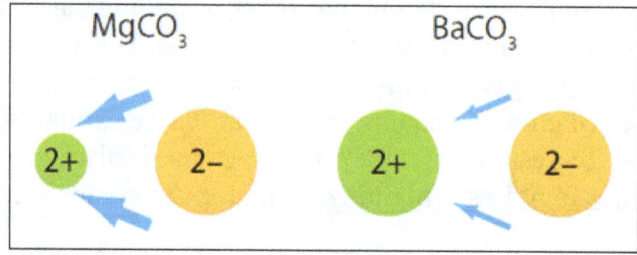

Magnesium's smaller 2+ ion (72 pm) polarises the carbonate 2- ion (135 pm) more than the larger barium 2+ ion (100 pm) because it has greater attractions for its electrons. The size of the 'electron drift' is indicated by the size of the blue arrows.

Diamond is an allotrope of carbon which was first confirmed in a very expensive experiment performed by the English chemist Smithson Tennant who was the first to burn a diamond in air (~600-800 °C) producing carbon dioxide.

Carbon dioxide as a non-metallic oxide is acidic. Carbon dioxide reacts with water in a reversible reaction to form carbonic acid (H_2CO_3).

$$CO_2(g) + H_2O(l) \rightleftharpoons H_2CO_3(aq)$$

The acid dissociation constant for carbonic acid $K_a = 4.45 \times 10^{-7}$ mol dm^{-3}. This indicates that it is a much weaker acid than ethanoic acid which has a $K_a = 1.74 \times 10^{-5}$ mol dm^{-3}. The equilibrium between carbon dioxide and water helps buffer the blood.

$$CO_2(aq) + H_2O(l) \rightleftharpoons HCO_3^-(aq) + H^+(aq)$$

The carbonic acid present in the equilibrium can neutralize hydroxide ions, which would increase the pH of the blood when added. The bicarbonate ion (HCO3-) can neutralize hydrogen ions (protons), which would cause a decrease in the pH of the blood when added. Both increasing and decreasing pH is life threatening.

CO_2 in Nature

The density of carbon dioxide is 1.53 g/cm³ at 21 °C. The fact carbon dioxide is heavier than air makes it particularly dangerous when large volumes are released from volcanoes. In 1986 a Cameroon volcano erupted suffocating 1700 villagers with some fortunate survivors asleep on the top bunk bed. When carbon dioxide gas is prepared it sinks to the bottom of a test tube or gas jar and will extinguish a flame. A very familiar test for carbon dioxide is its reaction with a colourless solution of calcium hydroxide which still retains its historical name lime water. A simple experiment is shown and the evolved carbon dioxide often described as effervescence (fizzing) will move from the left hand test tube into the right hand test tube.

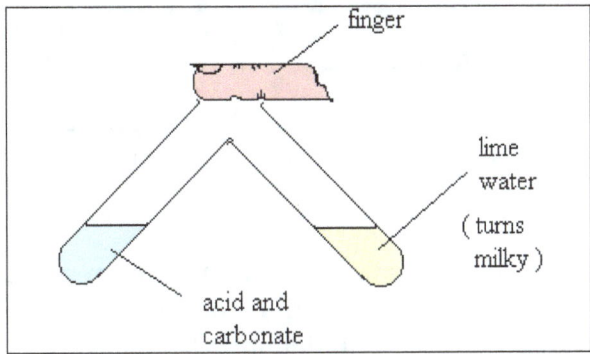

$$CO_2\,(g)\;+\;Ca(OH)_2\,(aq)\rightleftharpoons CaCO_3\,(s)\;+\;H_2O\,(l)$$

Lime, the fruit, tastes sour because it contains acids, whereas limewater (calcium hydroxide) is most definitely an alkali as it contains hydroxide (OH^-) ions.

The concentration of carbon dioxide in air is constantly being monitored by thousands of scientists across the globe. As a greenhouse gas it is 'believed' that changes in carbon dioxide levels contribute to increases in the earth's temperature, something which has not actually happened since 1997. One thing for certain is that the atmosphere on earth has evolved, and geologists are interested in how the gas became locked in carbonates.

Global Warming

The graph below shows how carbon dioxide levels have risen in the atmosphere over recent decades.

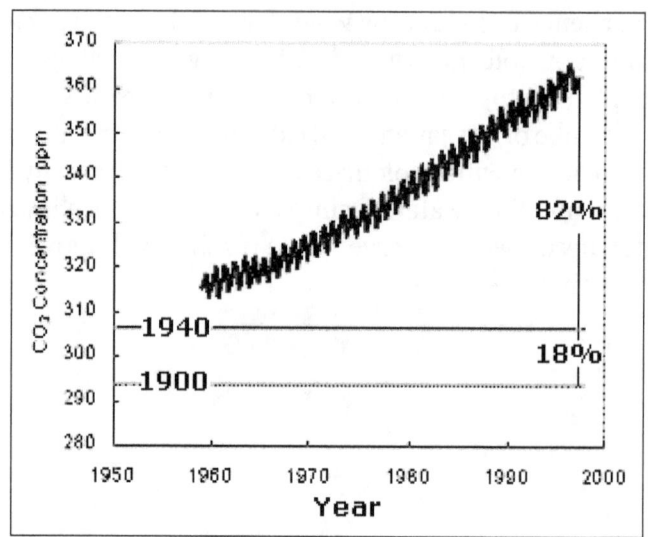

The intra-yearly variations can be attributed to the seasons and the presence of vast quantities of plankton and algae at certain times of the year. Carbon dioxide decreases during the growing season. There is no doubt that the amount of carbon dioxide in the atmosphere is increasing. What is in dispute is the effect of these increases on the Earth's temperature.

At the simplest level, photosynthesis can be represented as the equation which is the reverse of aerobic respiration. Photosynthesis takes place in green plants and fixes carbon dioxide into sugar molecules.

Photosynthesis: $6\,CO_2\,(g)\,+\,6\,H_2O\,(l) \rightarrow C_6H_{12}O_6\,(aq)\,+\,O_2\,(g)$

Gardeners have exploited this equation by increasing the concentration of carbon dioxide in greenhouses to improve yields. This can be as simple as lighting candles which are solid hydrocarbons (waxes) which when combusted produce carbon dioxide and water.

$$2\,C_{18}H_{38}\,(s)\,+\,37\,O_2\,(g) \rightarrow 36\,CO_2\,(g)\,+\,38\,H_2O\,(l)$$

The Carbon cycle is the mechanism by which the carbon dioxide levels are maintained in the atmosphere. Carbon dioxide is central to this important cycle.

A new technology called carbon capture and storage (CCS) will no doubt be added to future carbon cycles as it is a way to 'remove' carbon in the form of carbon dioxide from the air. Recent news has expressed concern about changes in pH and temperature of sea water and the impact this has on sensitive coral reefs around the world. Scientists often produce some of their best work when they mimic nature and this is what a company in the USA is achieving by mimicking corals by turning carbon dioxide into cement.

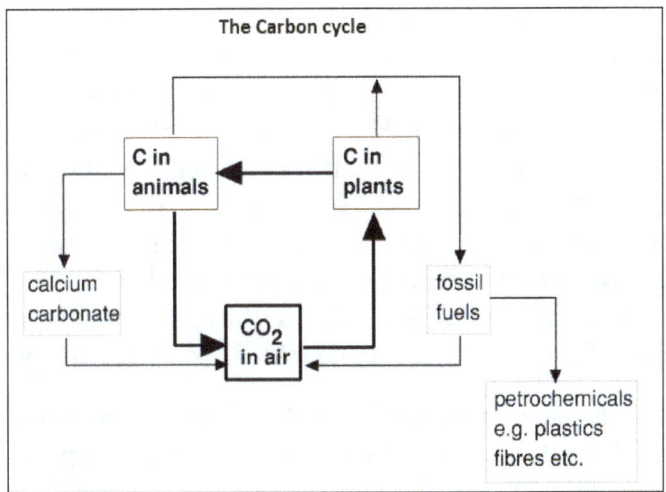

Carbon Capture (Methods to Remove Carbon Dioxide)

There is strong evidence that the Earth's temperature has warmed by 0.75°C since 1880. This increase coincides with rising atmospheric CO_2 levels. There has been a huge interest in the media about the link between this atmospheric gas and the enhanced greenhouse effect. It is now thought that CO2 levels have reached their highest in 15 to 20 million years (estimated at 389 parts per million in 2010). Carbon capture and storage (CCS) is a new technology that can prevent all the CO_2 being emitted primarily from pulverised coal combustion sites, into the atmosphere. Coal accounts for 40% of the world's carbon emissions. Instead, CO_2 can be 'captured' and compressed to a supercritical state (10 MPa) before being transported and stored from harm's way. Three main carbon capture methods are; post-combustion, pre-combustion and oxy-fuel combustion.

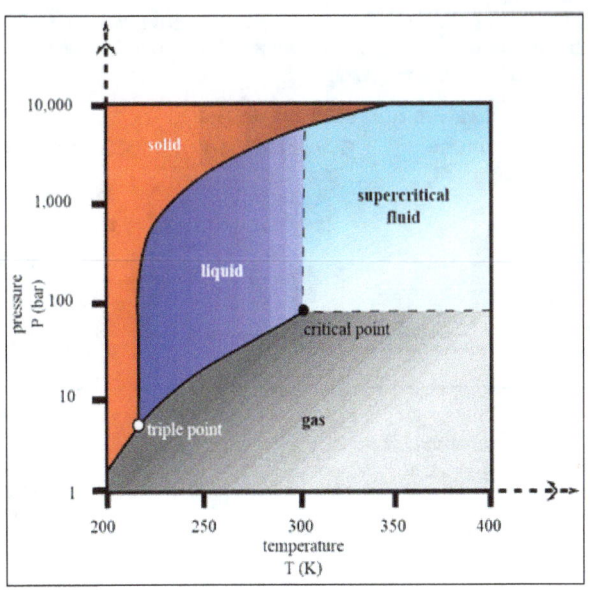

Post-combustion carbon capture occurs when CO_2 is extracted from flue gases after the fossil fuel is burned. Flue gases contain between 15 to 20% CO_2. The flue gases are passed through a solvent filter, often an amine, e.g. monoethanolamine ($H_2NCH_2CH_2OH$) or methyldiethanolamine ($CH_3N(C_2H_4OH)_2$), which can both absorb carbon dioxide. The amine solvent is then passed through a regenerator, which requires 80% of the total energy input for the process. Here, a change in pressure, and or, temperature is applied, releasing water vapour and leaving behind a concentrated stream of CO_2. The hydrate crystals formed from these flue gases hold a CO_2 concentration of 55-57%. This solvent can then be recycled back into the system and reused. Interestingly, the apparatus needed for this method can be retrofitted to existing power plants.

Pre-combustion carbon capture occurs before the fossil fuel is burned. This process is known as Integrated Gasification Combined Cycle (IGCC). The pulverised coal is first heated to between 537°C and 1426°C in pure oxygen in a 'gasifier'. This leads to the formation of synthesis gas (syngas) which is a mixture of carbon monoxide and hydrogen. CO is then passed through a shift reactor where steam is added, transforming CO into highly concentrated CO_2 and hydrogen. Finally, these gases are fed into a flask where they naturally begin to rise. An amine solution is then released into the flask and binds with the CO_2. With a newly increased density, this mixture falls to the bottom of the flask and is tapped off. This is now heated to 120°C in a new flask separating the mixture. CO_2 rises to the top for collection and the amine falls to the bottom for re-use. The remaining hydrogen can be used as a 'clean fuel' for heat, power and, potentially, as a transport fuel. This process can prevent approximately 90% of carbon emissions from entering the atmosphere. At the time of writing there are one hundred and sixty IGCC plants and another thirty-five IGCC plants in planning.

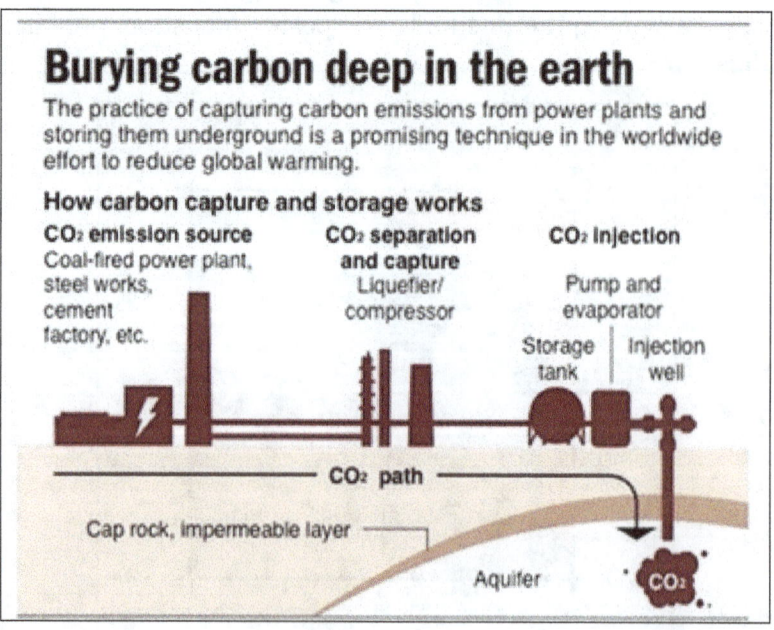

Burying carbon deep in the earth

The practice of capturing carbon emissions from power plants and storing them underground is a promising technique in the worldwide effort to reduce global warming.

How carbon capture and storage works

CO₂ emission source
Coal-fired power plant,
steel works,
cement
factory, etc.

CO₂ separation
and capture
Liquefier/
compressor

CO₂ Injection
Pump and
evaporator

Storage tank | Injection well

CO₂ path

Cap rock, impermeable layer

Aquifer   CO₂

The third carbon-capture process is oxy-fuel combustion. Here fossil fuel is burnt in a mixture of pure oxygen and recycled flue gas, which regulates the combustion temperature and cools down the reaction from 1550°C to 800°C. This results in more complete combustion, forming 90% CO_2 and water vapor. First, sulfur impurities are removed by adding limestone solution, which reacts with SO_2 to form gypsum (used in construction industry). Then the resulting mixture is cooled to allow the water vapor to condense and be tapped off. CO_2 is now compressed at 70 atm and transported to a storage unit. Before storage, captured carbon, in the form of CO_2, must be transported. The most common method of transportation is via pipelines typically made of carbon manganese steel. In the US alone, there are more than 1,500 miles of pipeline used for CO_2 transportation. CO_2 is mainly driven along pipelines in a dense-phase liquid state by a compressor. In this supercritical phase, CO_2 holds the density of a liquid and the viscosity of a gas.

In the final stage of carbon capture, the extracted CO_2, must be stored underground or underwater. In 2009, scientists mapped 6,000 m2 that could be used to store 500 years' worth of CO_2 emissions from the United States. Underground storage is commonly known as geological sequestration. Typically, CO_2 is injected 1,000 m below the ground into layers of porous sedimentary rock under an area described as 'cap rock'. This cap is impermeable and consequently acts as a seal. The pressure underground causes CO_2 to behave like a liquid. Declining oil fields and basalt rock formations are the two popular regions for storing CO_2. Approximately 30 to 50 million tons of CO_2 are injected into declining US oil fields every year. CO_2 captured in this way can be used for a process known as Enhanced Oil Recovery (EOR). Underground, at such high temperatures and pressures, injected CO_2 will dissolve in oil. This allows recovery of oil buried deep within porous rocks, which would otherwise be impossible to extract. If injected into basalt formations, eventually CO_2 is expected to react with available metal oxides to form stable carbonates, such as limestone. Their stability in most geological conditions precludes decomposition; and therefore little CO_2 is expected to escape back into the atmosphere.

Injecting CO_2 into the ocean at a depth below 11,482 metres is another safe potential method of CO_2 storage. The pressure at this depth compresses CO_2 into a 'slushy' material, which slowly sinks to the ocean floor. However, this method is as yet largely untested and estimates from a decade ago suggest it is likely to be prohibitively expensive at $40-80 per tonne of CO_2 stored.

Carbon Dioxide Detectors

CO_2 detectors are being used to detect dangerous levels of CO_2 in laboratories, breweries and cellars. This is important because levels of 1.5% CO_2 can cause headaches and hyperventilation, and levels exceeding 10% CO_2 levels can lead to death. Crowcon is a leading UK gas detection company that manufactures portable and fixed CO_2 monitors within enclosed area.

Carbon dioxide levels are detected using an infra-red detector. For ease of use the sensor continuously measures CO_2 levels and displays the reading on an easy-to-read backlit display. Units also provide a loud audio alarm. The detectors are manufactured by Crowcon, one of the UK's leading gas detection companies. The infra-red vibration modes for carbon dioxide are show. To be infra-red active a dipole is needed in this molecule which occurs when vibrations lead to asymmetry in the molecule, so dipoles do not cancel.

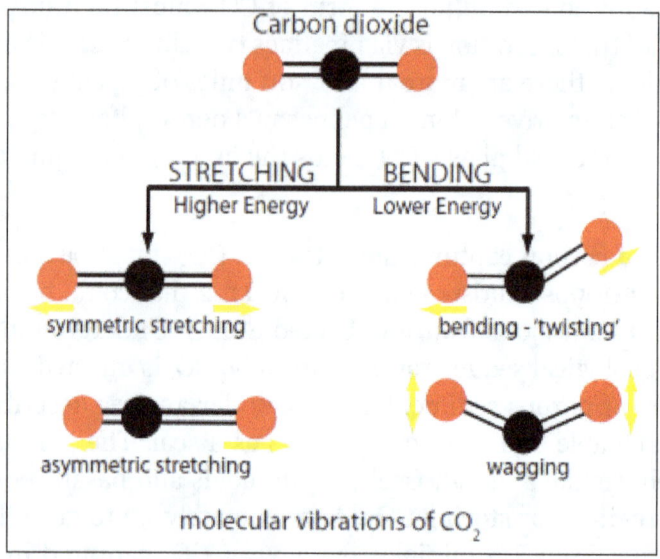

Carbon dioxide (CO_2) is a slightly toxic, odorless, colorless gas with a slightly pungent, acid taste. Carbon dioxide is a small but important constituent of air. It is a necessary raw material for most plant life, which remove carbon dioxide from air using the process of photosynthesis.

A typical concentration of CO_2 in air is currently about 0.040% or 404 ppm. The concentration of atmospheric carbon dioxide rises and falls in a seasonal pattern over a range of about 6 ppmv. The concentration of CO_2 in air has also been steadily increasing from year to year for over 70 years. The current rate of increase is about 2.5 ppm per year.

Carbon dioxide is formed by combustion and by biological processes. These include decomposition of organic material, fermentation and digestion. As an example, exhaled air contains as much as 4% carbon dioxide, or about 100 times the amount of carbon dioxide which was breathed in.

Large quantities of CO_2 are produced by lime kilns, which burn limestone (primarily calcium carbonate) to produce calcium oxide (lime, used to make cement); and in the production of magnesium from dolomite (calcium magnesium carbonate). Other industrial activities which produce large amounts of carbon dioxide are ammonia production and hydrogen production from natural gas or other hydrocarbon raw materials.

The concentration of CO_2 in air and in stack gases from simple combustion sources (heaters, boilers, furnaces) is not high enough to make carbon dioxide recovery commercially feasible. Producing carbon dioxide as a commercial product requires that it be recovered and purified from a relatively high-volume, CO_2-rich gas stream, generally a stream which is created as an unavoidable byproduct of a large-scale chemical production process or some form of biological process.

In almost all cases, carbon dioxide which is captured and purified for commercial applications would be vented to the atmosphere at the production point if it was not recovered for transport and beneficial use at other locations.

The most common operations from which commercially-produced carbon dioxide is recovered are industrial plants which produce hydrogen or ammonia from natural gas, coal, or other hydrocarbon feedstock, and large-volume fermentation operations in which plant products are made into ethanol for human consumption, automotive fuel, or industrial use. Breweries producing beer from various grain products are a traditional source. Corn-to-ethanol plants have been the most rapidly growing source of feed gas for CO_2 recovery. CO_2-rich natural gas reservoirs found in underground formations found primarily in the western United States and in Canada are another source of recoverable carbon dioxide.

CO_2 from both natural and industrial sources is used to enhance production of oil from older wells by injecting the carbon dioxide into appropriate underground formations. Carbon dioxide is used selectively, primarily in wells which will benefit not only from re-pressurization, but also from a reduction in viscosity of the oil in the reservoir caused by a portion of the CO_2 dissolving in the oil. (The extent to which carbon dioxide will dissolve in the oil varies with the type of petroleum present in the reservoir. If the viscosity reduction effect will be minimal, nitrogen, which is usually less expensive, may be used as the pressurant instead.)

Carbon dioxide will not burn or support combustion. Air with a carbon dioxide content of more than 10% will extinguish an open flame, and, if breathed, can be life-threatening. Such concentrations may build up in silos, digestion chambers, wells, sewers and the like. Caution must be exercised when entering these types of confined spaces.

CO_2 gas is 1.5 times as heavy as air, thus if released to the air it will concentrate at low elevations. Carbon dioxide will form "dry ice" at $-78.5°$ C ($-109.3°$ F). One kg of dry ice has the cooling capacity of 2 kg of ordinary ice. Gaseous or liquid carbon dioxide, stored under pressure, will form dry ice through an auto-refrigeration process if rapidly depressured.

Carbon dioxide is commercially available as high pressure cylinder gas, relatively low pressure (about 300 psig or 20 barg) refrigerated liquid, or as dry ice. Large quantities are produced and consumed at industrial sites making fertilizers, plastics and rubber.

Carbon dioxide is a versatile material, being used in many processes and applications

- each of which takes advantage of one or more these characteristics: reactivity, inertness and/ or coldness.

Carbon dioxide is commonly used as a raw material for production of various chemicals; as a working material in fire extinguishing systems; for carbonation of soft drinks; for freezing of food products such as poultry, meats, vegetables and fruit; for chilling of meats prior to grinding; for refrigeration and maintenance of ideal atmospheric conditions during transportation of food products to market; for enhancement of oil recovery from oil wells; and for treatment of alkaline water.

Carbon Dioxide in the Earth's Atmosphere

Carbon dioxide in air is considered to be a greenhouse gas because of its ability to absorb infrared light.

The concentration of CO_2 in the Earth's atmosphere has been increasing at a noticeable rate for much of the past century, There is much interest and concern over the inter-relationship between the levels of carbon dioxide in air and the subject of global warming.

Carbon dioxide plays a major role as a component of the carbon cycle in which carbon is exchanged between the atmosphere, the terrestrial biosphere (which includes freshwater systems and soil), the oceans, and sediments (including fossil fuels). These interactions are complex and widespread.

Carbon Dioxide (CO_2) Applications and Uses

Multi-Industry uses for Carbon Dioxide (CO_2)

Carbon dioxide in solid and in liquid form is used for refrigeration and cooling. It is used as an inert gas in chemical processes, in the storage of carbon powder and in fire extinguishers.

Metals Industry

Carbon dioxide is used in the manufacture of casting molds to enhance their hardness.

Manufacturing and Construction Uses

Carbon dioxide is used on a large scale as a shield gas in MIG/MAG welding, where the gas protects the weld puddle against oxidation by the surrounding air. A mixture of argon and carbon dioxide is commonly used today to achieve a higher welding rate and reduce the need for post weld treatment.

Dry ice pellets are used to replace sandblasting when removing paint from surfaces. It aids in reducing the cost of disposal and cleanup.

Chemicals, Pharmaceuticals and Petroleum Industry Uses

Large quantities are used as a raw material in the chemical process industry, especially for methanol and urea production.

Carbon dioxide is used in oil wells for oil extraction and to maintain pressure within a formation. When CO_2 is pumped into an oil well, it is partially dissolved into the oil, rendering it less viscous, allowing the oil to be extracted more easily from the bedrock. Considerably more oil can be extracted from through this process.

Rubber and Plastics Industry Uses

Flash is removed from rubber objects by tumbling them with crushed dry ice in a rotating drum.

Food and Beverages uses for Carbon Dioxide

Liquid or solid carbon dioxide is used for quick freezing, surface freezing, chilling and refrigeration in the transport of foods. In cryogenic tunnel and spiral freezers, high pressure liquid CO_2 is injected through nozzles that convert it to a mixture of CO_2 gas and dry ice "snow" that covers the surface of the food product. As it sublimates (goes directly from solid to gas states) refrigeration is transferred to the product.

Carbon dioxide gas is used to carbonate soft drinks, beers and wine and to prevent fungal and bacterial growth.

Liquid carbon dioxide is a good solvent for many organic compounds. It is used to de-caffeinate coffee.

It is used as an inert "blanket", as a product-dispensing propellant and an extraction agent. It can also be used to displace air during canning.

Supercritical CO_2 extraction coupled with a fractional separation technique is used by producers of flavors and fragrances to separate and purify volatile flavor and fragrances concentrates.

Cold sterilization can be carried out with a mixture of 90% carbon dioxide and 10% ethylene oxide, the carbon dioxide has a stabilizing effect on the ethylene oxide and reduces the risk of explosion.

Health Care Uses

Carbon dioxide is used as an additive to oxygen for medical use as a respiration stimulant.

Environmental Uses

Used as a propellant in aerosol cans, it replaces more environmentally troublesome alternatives.

By using dry ice pellets to replace sandblasting when removing paint from surfaces, problems of residue disposal are greatly reduced.

It is used to neutralize alkaline water.

Miscellaneous uses for Carbon Dioxide (CO_2)

Liquid carbon dioxide's solvent potential has been employed in some dry cleaning equipment as a substitute for conventional solvents. This use is still experimental - some types of soil are more effectively removed with traditional dry cleaning equipment, and the equipment is more expensive.

Yields of plant products grown in greenhouses can increase by 20% by enriching the air inside the greenhouse with carbon dioxide. The target level for enrichment is typically a carbon dioxide concentration of 1000 PPM (parts per million) - or about two and a half times the level present in the atmosphere.

OZONE

Although ozone was present at ground level before the Industrial Revolution, peak concentrations are now far higher than the pre-industrial levels, and even background concentrations well away from sources of pollution are substantially higher. Ozone acts as a greenhouse gas, absorbing some of the infrared energy emitted by the earth. Quantifying the greenhouse gas potency of ozone is difficult because it is not present in uniform concentrations across the globe. However, the most widely accepted scientific assessments relating to climate change (e.g. the Intergovernmental Panel on Climate Change Third Assessment Report) suggest that the radiative forcing of tropospheric ozone is about 25% that of carbon dioxide.

The annual global warming potential of tropospheric ozone is between 918–1022 tons carbon dioxide equivalent/tons tropospheric ozone. This means on a per-molecule basis, ozone in the troposphere has a radiative forcing effect roughly 1,000 times as strong as carbon dioxide. However, tropospheric ozone is a short-lived greenhouse gas, which decays in the atmosphere much more quickly than carbon dioxide. This means that over a 20-year span, the global warming potential of tropospheric ozone is much less, roughly 62 to 69 tons carbon dioxide equivalent/ton tropospheric ozone.

Because of its short-lived nature, tropospheric ozone does not have strong global effects, but has very strong radiative forcing effects on regional scales. In fact, there are regions of the world where tropospheric ozone has a radiative forcing up to 150% of carbon dioxide.

Health Effects

For the last few decades, scientists studied the effects of acute and chronic ozone exposure on human health. Hundreds of studies suggest that ozone is harmful to people at levels currently found in urban areas. Ozone has been shown to affect the respiratory, cardiovascular and central nervous system. Early death and problems in reproductive health and development are also shown to be associated with ozone exposure.

Vulnerable Populations

The American Lung Association has identified five populations who are especially vulnerable to the effects of breathing ozone:

- Children and teens.

- People 65 years old and older.

- People who work or exercise outdoors.

- People with existing lung diseases, such as asthma and chronic obstructive pulmonary disease (also known as COPD, which includes emphysema and chronic bronchitis).

- People with cardiovascular disease.

Additional evidence suggests that women, those with obesity and low-income populations may also face higher risk from ozone although more research is needed.

Acute Ozone Exposure

Acute ozone exposure ranges from hours to a few days. Because ozone is gas, it causes direct and immediate harm to the lungs and the entire respiratory system. Inhaled ozone causes acute but reversible changes in lung function and inflammation, as well as airway hyperresponsiveness. These changes lead to shortness of breath, wheezing, and coughing which may exacerbate lung diseases, like asthma or chronic obstructive pulmonary disease (COPD) resulting in the need to receive medical treatment. Acute and chronic exposure to ozone has been shown to cause an increased risk of respiratory infections, due to the following mechanism.

Multiple studies have been conducted to determine the mechanism behind ozone's harmful effects, particularly in the lungs. These studies have shown that exposure to ozone causes changes in the immune response within the lung tissue, resulting in disruption of both the innate and adaptive immune response, as well as altering the protective function of lung epithelial cells. It is thought that these changes in immune response and the related inflammatory response are factors that likely contribute to the increased risk of lung infections, and worsening or triggering of asthma and reactive airways after exposure to ground-level ozone pollution.

The innate (cellular) immune system consists of various chemical signals and cell types that work broadly and against multiple pathogen types, typically bacteria or foreign bodies/substances in the host. The cells of the innate system include phagocytes, neutrophils, both thought to contribute to the mechanism of ozone pathology in the lungs, as the functioning of these cell types have been shown to change after exposure to ozone. Macrophages, cells that serve the purpose of eliminating pathogens or foreign material through the process of "phagocytosis", have been shown to change the level of inflammatory signals they release in response to ozone, either up-regulating and resulting in an inflammatory response in the lung, or down-regulating and reducing immune protection. Neutrophils, another important cell type of the innate immune system that primarily targets bacterial pathogens, are found to be present in the airways within 6 hours of exposure to high ozone levels. Despite high levels in the lung tissues, however, their ability to clear bacteria appears impaired by exposure to ozone.

The adaptive immune system is the branch of immunity that provides long-term protection via the development of antibodies targeting specific pathogens and is also impacted by high ozone exposure. Lymphocytes, a cellular component of the adaptive immune response, produce an increased amount of inflammatory chemicals called "cytokines" after exposure to ozone, which may contribute to airway hyperreactivity and worsening asthma symptoms.

The airway epithelial cells also play an important role in protecting individuals from pathogens. In normal tissue, the epithelial layer forms a protective barrier, and also contains specialized ciliary structures that work to clear foreign bodies, mucus and pathogens from the lungs. When exposed to ozone, the cilia become damaged and mucociliary clearance of pathogens is reduced. Furthermore, the epithelial barrier becomes weakened, allowing pathogens to cross the barrier, proliferate and spread into deeper tissues. Together, these changes in the epithelial barrier help make individuals more susceptible to pulmonary infections.

Inhaling ozone not only affects the immune system and lungs, but it may also affect the heart as well. Ozone causes short-term autonomic imbalance leading to changes in heart rate and reduction in heart rate variability; and high levels exposure for as little as one-hour results in a supraventricular arrhythmia in the elderly, both increase the risk of premature death and stroke. Ozone may also lead to vasoconstriction resulting in increased systemic arterial pressure contributing to increased risk of cardiac morbidity and mortality in patients with pre-existing cardiac diseases.

Chronic Ozone Exposure

Breathing ozone for periods longer than eight hours at a time for weeks, months or years defines chronic exposure. Numerous studies suggest a serious impact on the health of various populations from this exposure.

One study finds significant positive associations between chronic ozone and all-cause, circulatory, and respiratory mortality with 2%, 3%, and 12% increases in risk per 10 ppb and report an association (95% CI) of annual ozone and all-cause mortality with a hazard ratio of 1.02 (1.01–1.04), and with cardiovascular mortality of 1.03 (1.01–1.05). Adding to an additional study, which suggests similar associations with all-cause mortality and even larger effects for cardiovascular mortality.

Chronic ozone has detrimental effects on children, especially those with asthma. The risk for hospitalization in children with asthma increases with chronic exposure to ozone; younger children and those with low-income status are even at greater risk.

Adults suffering from respiratory diseases (asthma, COPD, lung cancer) are at a higher risk of mortality and morbidity and critically ill patients have an increased risk of developing acute respiratory distress syndrome with chronic ozone exposure as well.

Ozone Air Pollution

Red Alder leaf, showing discolouration caused by ozone pollution.

Signboard in Gulfton, Houston indicating an ozone watch.

Ozone precursors are a group of pollutants, predominantly those emitted during the combustion of fossil fuels. Ground-level ozone pollution (tropospheric ozone) is created near the Earth's surface by the action of daylight UV rays on these precursors. The ozone at ground level is primarily from fossil fuel precursors, but methane is a natural precursor, and the very low natural background level of ozone at ground level is considered safe.

There is a great deal of evidence to show that ground-level ozone can harm lung function and irritate the respiratory system. Exposure to ozone (and the pollutants that produce it) is linked to premature death, asthma, bronchitis, heart attack, and other cardiopulmonary problems.

Long-term exposure to ozone has been shown to increase risk of death from respiratory illness. A study of 450,000 people living in United States cities saw a significant correlation between ozone levels and respiratory illness over the 18-year follow-up period. The study revealed that people living in cities with high ozone levels, such as Houston or Los Angeles, had an over 30% increased risk of dying from lung disease.

Air quality guidelines such as those from the World Health Organization, the United States Environmental Protection Agency (EPA) and the European Union are based on detailed studies designed to identify the levels that can cause measurable ill health effects.

According to scientists with the US EPA, susceptible people can be adversely affected by ozone levels as low as 40 nmol/mol. In the EU, the current target value for ozone concentrations is 120 $\mu g/m^3$ which is about 60 nmol/mol. This target applies to all member states in accordance with Directive 2008/50/EC. Ozone concentration is measured as a maximum daily mean of 8 hour averages and the target should not be exceeded on more than 25 calendar days per year, starting from January 2010. Whilst the directive requires in the future a strict compliance with 120 $\mu g/m^3$ limit (i.e. mean ozone concentration not to be exceeded on any day of the year), there is no date set for this requirement and this is treated as a long-term objective.

In the USA, the Clean Air Act directs the EPA to set National Ambient Air Quality Standards for several pollutants, including ground-level ozone, and counties out of compliance with these standards are required to take steps to reduce their levels. In May 2008, under a court order, the EPA lowered its ozone standard from 80 nmol/mol to 75 nmol/mol. The move proved controversial, since the Agency's own scientists and advisory board had recommended lowering the standard to 60 nmol/mol. Many public health and environmental groups also supported the 60 nmol/mol standard, and the World Health Organization recommends 51 nmol/mol.

On January 7, 2010, the U.S. Environmental Protection Agency (EPA) announced proposed revisions to the National Ambient Air Quality Standard (NAAQS) for the pollutant ozone, the principal component of smog:

> EPA proposes that the level of the 8-hour primary standard, which was set at 0.075 $\mu mol/mol$ in the 2008 final rule, should instead be set at a lower level within the range of 0.060 to 0.070 $\mu mol/mol$, to provide increased protection for children and other at risk populations against an array of O_3 – related adverse health effects that range from decreased lung function and increased respiratory symptoms to serious indicators of respiratory morbidity including emergency department visits and hospital admissions for respiratory causes, and possibly cardiovascular-related morbidity as well as total non- accidental and cardiopulmonary mortality.

On October 26, 2015, the EPA published a final rule with an effective date of December 28, 2015 that revised the 8-hour primary NAAQS from 0.075 ppm to 0.070 ppm.

The EPA has developed an air quality index (AQI) to help explain air pollution levels to the general public. Under the current standards, eight-hour average ozone mole fractions of 85 to 104 nmol/mol are described as "unhealthy for sensitive groups", 105 nmol/mol to 124 nmol/mol as "unhealthy", and 125 nmol/mol to 404 nmol/mol as "very unhealthy".

Ozone can also be present in indoor air pollution, partly as a result of electronic equipment such as photocopiers. A connection has also been known to exist between the increased pollen, fungal spores, and ozone caused by thunderstorms and hospital admissions of asthma sufferers.

In the Victorian era, one British folk myth held that the smell of the sea was caused by ozone. In fact, the characteristic "smell of the sea" is caused by dimethyl sulfide, a chemical generated by phytoplankton. Victorian Britons considered the resulting smell "bracing".

Heat Waves

An investigation to assess the joint effects of ozone and heat during the European heat waves in 2003, concluded that these appear to be additive.

Physiology

Ozone, along with reactive forms of oxygen such as superoxide, singlet oxygen, hydrogen peroxide, and hypochlorite ions, is produced by white blood cells and other biological systems (such as the roots of marigolds) as a means of destroying foreign bodies. Ozone reacts directly with organic double bonds. Also, when ozone breaks down to dioxygen it gives rise to oxygen free radicals, which are highly reactive and capable of damaging many organic molecules. Moreover, it is believed that the powerful oxidizing properties of ozone may be a contributing factor of inflammation. The cause-and-effect relationship of how the ozone is created in the body and what it does is still under consideration and still subject to various interpretations, since other body chemical processes can trigger some of the same reactions. A team headed by Paul Wentworth Jr. of the Department of Chemistry at the Scripps Research Institute has shown evidence linking the antibody-catalyzed water-oxidation pathway of the human immune response to the production of ozone. In this system, ozone is produced by antibody-catalyzed production of trioxidane from water and neutrophil-produced singlet oxygen.

When inhaled, ozone reacts with compounds lining the lungs to form specific, cholesterol-derived metabolites that are thought to facilitate the build-up and pathogenesis of atherosclerotic plaques (a form of heart disease). These metabolites have been confirmed as naturally occurring in human atherosclerotic arteries and are categorized into a class of secosterols termed atheronals, generated by ozonolysis of cholesterol's double bond to form a 5,6 secosterol as well as a secondary condensation product via aldolization.

Ozone has been implicated to have an adverse effect on plant growth: "ozone reduced total chlorophylls, carotenoid and carbohydrate concentration, and increased 1-aminocyclopropane-1-carboxylic acid (ACC) content and ethylene production. In treated plants, the ascorbate leaf pool was decreased, while lipid peroxidation and solute leakage were significantly higher than in ozone-free controls. The data indicated that ozone triggered protective mechanisms against oxidative stress in citrus." Studies that have used pepper plants as a model have shown that ozone decreased fruit yield and changed fruit quality. Furthermore, it was also observed a decrease in chlorophylls levels and antioxidant defences on the leaves, as well as increased the reactive oxygen species (ROS) levels and lipid and protein damages.

Safety Regulations

Because of the strongly oxidizing properties of ozone, ozone is a primary irritant, affecting especially the eyes and respiratory systems and can be hazardous at even low concentrations. The Canadian Centre for Occupation Safety and Health reports that:

> "Even very low concentrations of ozone can be harmful to the upper respiratory tract and the lungs. The severity of injury depends on both by the concentration of ozone and the duration of exposure. Severe and permanent lung injury or death could result from even a very short-term exposure to relatively low concentrations."

To protect workers potentially exposed to ozone, U.S. Occupational Safety and Health Administration has established a permissible exposure limit (PEL) of 0.1 μmol/mol (29 CFR 1910.1000 table Z-1), calculated as an 8-hour time weighted average. Higher concentrations are especially hazardous and NIOSH has established an Immediately Dangerous to Life and Health Limit (IDLH) of 5 μmol/mol. Work environments where ozone is used or where it is likely to be produced should have adequate ventilation and it is prudent to have a monitor for ozone that will alarm if the concentration exceeds the OSHA PEL. Continuous monitors for ozone are available from several suppliers.

Elevated ozone exposure can occur on passenger aircraft, with levels depending on altitude and atmospheric turbulence. United States Federal Aviation Authority regulations set a limit of 250 nmol/mol with a maximum four-hour average of 100 nmol/mol. Some planes are equipped with ozone converters in the ventilation system to reduce passenger exposure.

Production

Ozone production demonstration, Fixed Nitrogen Research Laboratory, 1926.

Ozone generators are used to produce ozone for cleaning air or removing smoke odours in unoccupied rooms. These ozone generators can produce over 3 g of ozone per hour. Ozone often forms in nature under conditions where O_2 will not react. Ozone used in industry is measured in μmol/mol (ppm, parts per million), nmol/mol (ppb, parts per billion), μg/m³, mg/h (milligrams per hour) or weight percent. The regime of applied concentrations ranges from 1% to 5% (in air) and from 6% to 14% (in oxygen) for older generation methods. New electrolytic methods can achieve up 20% to 30% dissolved ozone concentrations in output water.

Temperature and humidity play a large role in how much ozone is being produced using traditional generation methods (such as corona discharge and ultraviolet light). Old generation methods will produce less than 50% of nominal capacity if operated with humid ambient air, as opposed to very dry air. New generators, using electrolytic methods, can achieve higher purity and dissolution through using water molecules as the source of ozone production.

Corona Discharge Method

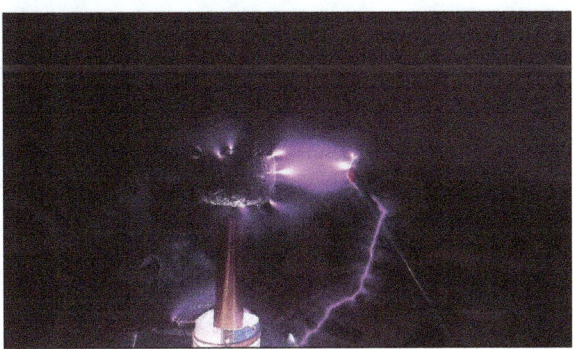

A homemade ozone generator. Ozone is produced in the corona discharge.

This is the most common type of ozone generator for most industrial and personal uses. While variations of the "hot spark" coronal discharge method of ozone production exist, including medical grade and industrial grade ozone generators, these units usually work by means of a corona discharge tube. They are typically cost-effective and do not require an oxygen source other than the ambient air to produce ozone concentrations of 3–6%. Fluctuations in ambient air, due to weather or other environmental conditions, cause variability in ozone production. However, they also produce nitrogen oxides as a by-product. Use of an air dryer can reduce or eliminate nitric acid formation by removing water vapor and increase ozone production. Use of an oxygen concentrator can further increase the ozone production and further reduce the risk of nitric acid formation by removing not only the water vapor, but also the bulk of the nitrogen.

Ultraviolet Light

UV ozone generators, or vacuum-ultraviolet (VUV) ozone generators, employ a light source that generates a narrow-band ultraviolet light, a subset of that produced by the Sun. The Sun's UV sustains the ozone layer in the stratosphere of Earth.

UV ozone generators use ambient air for ozone production, no air prep systems are used (air dryer or oxygen concentrator), therefore these generators tend to be less expensive. However UV ozone generators usually produce ozone with a concentration of about 0.5% or lower which limits the potential ozone production rate. Another disadvantage of this method is that it requires the ambient air (oxygen) to be exposed to the UV source for a longer amount of time, and any gas that is not exposed to the UV source will not be treated. This makes UV generators impractical for use in situations that deal with rapidly moving air or water streams (in-duct air sterilization, for example). Production of ozone is one of the potential dangers of ultraviolet germicidal irradiation. VUV ozone generators are used in swimming pool and spa applications ranging to millions of gallons of water. VUV ozone generators, unlike corona discharge generators, do not produce harmful nitrogen by-products and also unlike corona discharge systems, VUV ozone generators work extremely well in humid air environments. There is also not normally a need for expensive off-gas mechanisms, and no need for air driers or oxygen concentrators which require extra costs and maintenance.

Cold Plasma

In the cold plasma method, pure oxygen gas is exposed to a plasma created by dielectric barrier discharge. The diatomic oxygen is split into single atoms, which then recombine in triplets to form ozone.

Cold plasma machines utilize pure oxygen as the input source and produce a maximum concentration of about 5% ozone. They produce far greater quantities of ozone in a given space of time compared to ultraviolet production. However, because cold plasma ozone generators are very expensive, they are found less frequently than the previous two types.

The discharges manifest as filamentary transfer of electrons (micro discharges) in a gap between two electrodes. In order to evenly distribute the micro discharges, a dielectric insulator must be used to separate the metallic electrodes and to prevent arcing.

Some cold plasma units also have the capability of producing short-lived allotropes of oxygen which include O_4, O_5, O_6, O_7, etc. These species are even more reactive than ordinary O_3.

Electrolytic

Electrolytic ozone generation (EOG) splits water molecules into H_2, O_2, and O_3. In most EOG methods, the hydrogen gas will be removed to leave oxygen and ozone as the only reaction products. Therefore, EOG can achieve higher dissolution in water without other competing gases found in corona discharge method, such as nitrogen gases present in ambient air. This method of generation can achieve concentrations of 20–30% and is independent of air quality because water is used as the source material. Production of ozone electrolytically is typically unfavorable because of the high overpotential required to produce ozone as compared to oxygen. This is why ozone is not produced during typical water electrolysis. However, it is possible to increase the overpotential of oxygen by careful catalyst selection such that ozone is preferentially produced under electrolysis. Catalysts typically chosen for this approach are lead dioxide or boron-doped diamond.

The ozone to oxygen ratio is improved by increasing current density at the anode, cooling the electrolyte around the anode close to 0°C, using an acidic electrolyte (such as dilute sulfuric acid) instead of a basic solution, and by applying pulsed current instead of DC.

Special Considerations

Ozone cannot be stored and transported like other industrial gases (because it quickly decays into diatomic oxygen) and must therefore be produced on site. Available ozone generators vary in the arrangement and design of the high-voltage electrodes. At production capacities higher than 20 kg per hour, a gas/water tube heat-exchanger may be utilized as ground electrode and assembled with tubular high-voltage electrodes on the gas-side. The regime of typical gas pressures is around 2 bars (200 kPa) absolute in oxygen and 3 bars (300 kPa) absolute in air. Several megawatts of electrical power may be installed in large facilities, applied as one phase AC current at 50 to 8000 Hz and peak voltages between 3,000 and 20,000 volts. Applied voltage is usually inversely related to the applied frequency.

The dominating parameter influencing ozone generation efficiency is the gas temperature, which is controlled by cooling water temperature and gas velocity. The cooler the water, the better the ozone synthesis. The lower the gas velocity, the higher the concentration (but the lower the net ozone produced). At typical industrial conditions, almost 90% of the effective power is dissipated as heat and needs to be removed by a sufficient cooling water flow.

Because of the high reactivity of ozone, only a few materials may be used like stainless steel (quality 316L), titanium, aluminium (as long as no moisture is present), glass, polytetrafluorethylene, or polyvinylidene fluoride. Viton may be used with the restriction of constant mechanical forces and absence of humidity (humidity limitations apply depending on the formulation). Hypalon may be used with the restriction that no water come in contact with it, except for normal atmospheric levels. Embrittlement or shrinkage is the common mode of failure of elastomers with exposure to ozone. Ozone cracking is the common mode of failure of elastomer seals like O-rings.

Silicone rubbers are usually adequate for use as gaskets in ozone concentrations below 1 wt%, such as in equipment for accelerated aging of rubber samples.

Incidental Production

Ozone may be formed from O_2 by electrical discharges and by action of high energy electromagnetic radiation. Unsuppressed arcing in electrical contacts, motor brushes, or mechanical switches breaks down the chemical bonds of the atmospheric oxygen surrounding the contacts $[O_2 \rightarrow 2O]$. Free radicals of oxygen in and around the arc recombine to create ozone $[O_3]$. Certain electrical equipment generate significant levels of ozone. This is especially true of devices using high voltages, such as ionic air purifiers, laser printers, photocopiers, tasers and arc welders. Electric motors using brushes can generate ozone from repeated sparking inside the unit. Large motors that use brushes, such as those used by elevators or hydraulic pumps, will generate more ozone than smaller motors.

Ozone is similarly formed in the Catatumbo lightning storms phenomenon on the Catatumbo River in Venezuela, though ozone's instability makes it dubious that it has any effect on the ozonosphere. It is the world's largest single natural generator of ozone, lending calls for it to be designated a UNESCO World Heritage Site.

Laboratory Production

In the laboratory, ozone can be produced by electrolysis using a 9 volt battery, a pencil graphite rod cathode, a platinum wire anode and a 3 molar sulfuric acid electrolyte. The half cell reactions taking place are:

$$3\ H_2O \rightarrow O_3 + 6\ H^+ + 6\ e^-\ (\Delta E^\circ = -1.53\ V)$$

$$6 \text{ H}^+ + 6 \text{ e}^- \rightarrow 3 \text{ H}_2 \ (\Delta E^\circ = 0 \text{ V})$$

$$2 \text{ H}_2\text{O} \rightarrow \text{O}_2 + 4 \text{ H}^+ + 4 \text{ e}^- \ (\Delta E^\circ = 1.23 \text{ V})$$

In the net reaction, three equivalents of water are converted into one equivalent of ozone and three equivalents of hydrogen. Oxygen formation is a competing reaction.

It can also be generated by a high voltage arc. In its simplest form, high voltage AC, such as the output of a neon-sign transformer is connected to two metal rods with the ends placed sufficiently close to each other to allow an arc. The resulting arc will convert atmospheric oxygen to ozone.

It is often desirable to contain the ozone. This can be done with an apparatus consisting of two concentric glass tubes sealed together at the top with gas ports at the top and bottom of the outer tube. The inner core should have a length of metal foil inserted into it connected to one side of the power source. The other side of the power source should be connected to another piece of foil wrapped around the outer tube. A source of dry O_2 is applied to the bottom port. When high voltage is applied to the foil leads, electricity will discharge between the dry dioxygen in the middle and form O_3 and O_2 which will flow out the top port. This is called a Siemen's ozoniser. The reaction can be summarized as follows:

$$3O_2 \xrightarrow{\ electricity\ } 2O_3$$

CARBON MONOXIDE

Carbon monoxide (CO) is only a very weak direct greenhouse gas, but has important indirect effects on global warming. Carbon monoxide reacts with hydroxyl (OH) radicals in the atmosphere, reducing their abundance. As OH radicals help to reduce the lifetimes of strong greenhouse gases, like methane, carbon monoxide indirectly increases the global warming potential of these gases.

Carbon monoxide in the atmosphere can also lead to the formation of the tropospheric greenhouse gas 'ozone'. Atmospheric concentrations of carbon monoxide vary widely around the world and throughout the year, ranging from as low as 30 parts per billion up to around 200 parts per billion. Concentrations increased during the 20th century, but there are some signs that concentrations dropped slightly in the 1990s due to widespread use of catalytic convertors, with their lower carbon monoxide emissions, in cars.

Aside from man-made sources, a great deal of carbon monoxide comes from the chemical oxidation of methane and other hydrocarbons in our atmosphere. Additional natural sources include emission from vegetation and the world's oceans. By far the largest sink for carbon monoxide is its reaction with OH in the atmosphere, as noted previously. However, a small but significant amount is also lost from the atmosphere through deposition on the ground.

Human Impact

Today more than half of carbon monoxide emissions are man-made. The highest concentrations of carbon monoxide tend to occur close to areas of high human population. On a global scale, this has meant that the more densely populated northern hemisphere has higher concentrations of carbon monoxide than the southern hemisphere. Biomass burning and fossil fuel use are the main sources of man-made carbon monoxide emissions.

Potential for Control

As with many direct and indirect greenhouse gases, reductions in carbon monoxide emissions can most effectively be made through direct reductions in fossil fuel use. There is some evidence that the widespread use of catalytic convertors in cars has significantly reduced carbon monoxide emissions from this source. However, such reductions must be balanced against the increased emissions of the greenhouse gases carbon dioxide and nitrous oxide which often result from a switch to catalytic converters.

METHANE

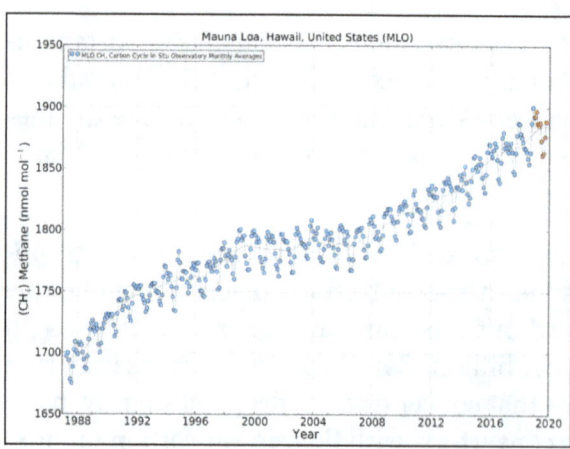

Methane concentrations up to October 2019: A monthly peak of 1900.49 ppb was reached in November 2018.

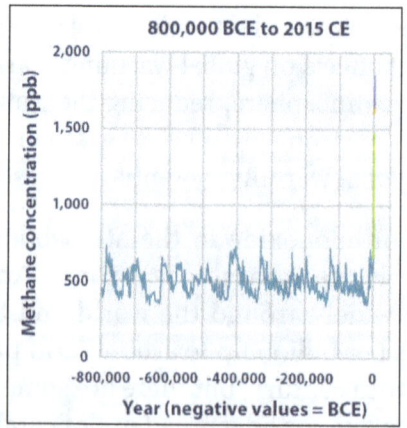

Compilation of paleo-climatology data of methane.

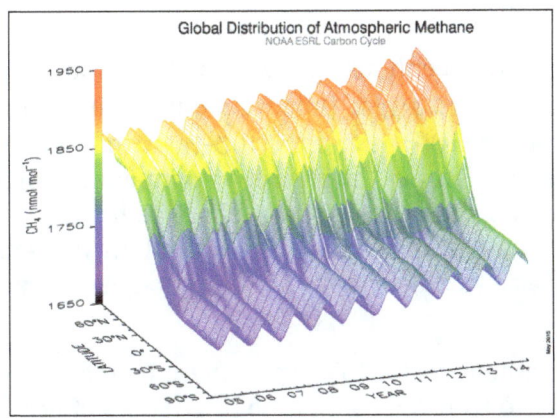

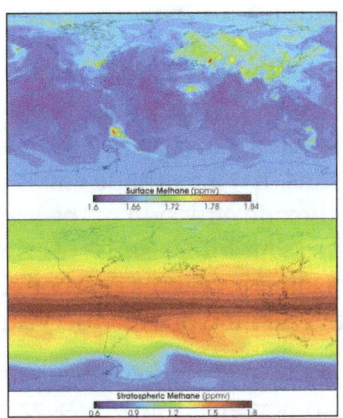

Methane observations from 2005 to 2014 showing the seasonal variations and the difference between northern and southern hemispheres.

Computer models showing the amount of methane (parts per million by volume) at the surface (top) and in the stratosphere (bottom).

Atmospheric methane is the methane present in Earth's atmosphere. Atmospheric methane concentrations are of interest because it is one of the most potent greenhouse gases in Earth's atmosphere. Atmospheric methane is rising.

The 20-year global warming potential of methane is 84. That is, over a 20-year period, it traps 84 times more heat per mass unit than carbon dioxide and 32 times the effect when accounting for aerosol interactions. Global methane concentrations rose from 722 parts per billion (ppb) in pre-industrial times to 1866 ppb by 2019, an increase by a factor of 2.5 and the highest value in at least 800,000 years. Its concentration is higher in the Northern Hemisphere since most sources (both natural and human) are located on land and the Northern Hemisphere has more land mass. The concentrations vary seasonally, with, for example, a minimum in the northern tropics during April–May mainly due to removal by the hydroxyl radical.

Early in the Earth's history carbon dioxide and methane likely produced a greenhouse effect. The carbon dioxide would have been produced by volcanoes and the methane by early microbes. During this time, Earth's earliest life appeared. These first, ancient bacteria added to the methane concentration by converting hydrogen and carbon dioxide into methane and water. Oxygen did not become a major part of the atmosphere until photosynthetic organisms evolved later in Earth's history. With no oxygen, methane stayed in the atmosphere longer and at higher concentrations than it does today.

The known sources of methane are predominantly located near the Earth's surface. In combination with vertical atmospheric motions and methane's relatively long lifetime, methane is considered to be a well-mixed gas. In other words, the concentration of methane is taken to be constant with respect to height within the troposphere. The dominant sink of methane in the troposphere is reaction with hydroxyl radicals that are formed by reaction of singlet oxygen atoms with water vapor. Methane is also present in the stratosphere, where methane's concentration decreases with height.

Methane as a Greenhouse Gas

Methane in the Earth's atmosphere is a strong greenhouse gas with a global warming potential (GWP) 104 times greater than CO_2 in a 20-year time frame; methane is not as persistent a gas as CO_2 and tails off to about GWP of 28 for a 100-year time frame. This means that a methane emission will have 28 times the impact on temperature of a carbon dioxide emission of the same mass over the following 100 years. Methane has a large effect but for a relatively brief period, having an estimated lifetime of 9.1 years in the atmosphere, whereas carbon dioxide has a small effect for a long period, having an estimated lifetime of over 100 years.

The globally averaged concentration of methane in Earth's atmosphere increased by about 150 percent from 722 ± 25 ppb in 1750 to 1803.2 ± 1.2 ppb in 2011. As of 2011, methane contributed radiative forcing of 0.48 ± 0.05 Wm^{-2}, or about 17% of the total radiative forcing from all of the long-lived and globally mixed greenhouse gases. According to NOAA, the atmospheric methane concentration has continued to increase since 2011 to an average global concentration of 1850.5 ppb as of July 2018. The May 2018 peak was 1854.8 ppb, while the May 2019 peak was 1862.8 ppb, a .3% increase.

Global Methane Cycle

This simple diagram depicts the flow of methane from sources into the atmosphere as well as the sinks that consume methane. More detailed explanations of each source and sink are covered in later sections.

- Permafrost, glaciers, and ice cores: A source that slowly releases methane trapped in frozen environments as global temperatures rise.

- Wetlands: Warm temperatures and moist environments are ideal for methane production.

- Forest fire: Mass burning of organic matter releases methane into the atmosphere.

- Rice paddies: The warmer and moister the rice field, the more methane is produced.

- Animals: Microorganisms, breaking down difficult to digest material in the guts of ruminant livestock and termites, produce methane that is then released during defecation.

- Plants: While methane can be consumed in soil before being released into the atmosphere, plants allow for direct travel of methane up through the roots and leaves and into the atmosphere. Plants may also be direct producers of methane.

- Landfills: Decaying organic matter and anaerobic conditions cause landfills to be a significant source of methane.

- Waste water treatment facilities: Anaerobic treatment of organic compounds in the water results in the production of methane.

- Hydroxyl radical: OH in the atmosphere is the largest sink for atmospheric methane as well as one of the most significant sources of water vapor in the upper atmosphere.

- Chlorine radical: Free chlorine in the atmosphere also reacts with methane.

Other sources of methane include:

- Natural gas extraction, transportation and use, hydraulic fracturing.

- Natural gas seeps from coal areas and natural gas deposits.

- Methane hydrates located around the world on the seafloor.

Emissions Accounting of Methane

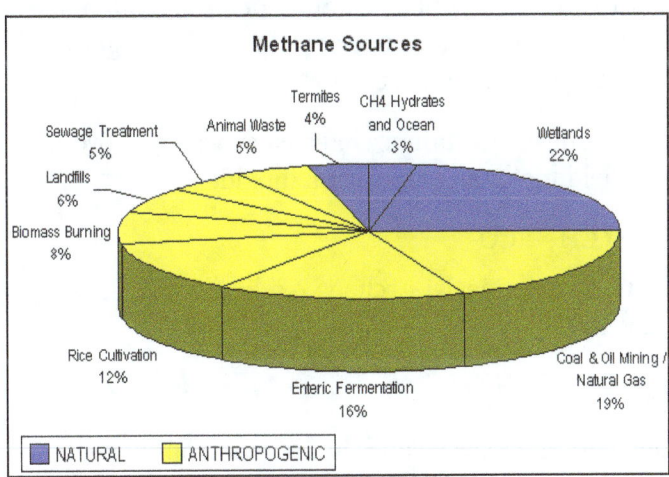

Natural and anthropogenic methane sources, according to the
NASA Goddard Institute for Space Studies.

The balance between sources and sinks of methane is not yet fully understood. The IPCC Working Group I stated in chapter 2 of the Fourth Assessment Report that there are "large uncertainties in the current bottom-up estimates of components of the global source", and the balance between sources and sinks is not yet well known. The most important sink in the methane cycle is reaction with the hydroxyl radical, which is produced photochemically in the atmosphere. Production of this radical is not fully understood and has a large effect on atmospheric concentrations. This uncertainty is exemplified by observations that have shown between the year 2000 and 2006 increases in atmospheric concentration of methane ceased, for reasons still being investigated.

Natural Sources of Atmospheric Methane

Any process that results in the production of methane and its release into the atmosphere can be considered a "source." The two main processes that are responsible for methane production occur as a result of microorganisms anaerobically converting organic compounds into methane.

Methanogenesis

Most ecological emissions of methane relate directly to methanogens generating methane in warm, moist soils as well as in the digestive tracts of certain animals. Methanogens are methane producing microorganisms. In order to produce energy, they use an anaerobic process called methanogenesis. This process is used in lieu of aerobic, or with oxygen, processes because methanogens are unable to metabolise in the presence of even small concentrations of oxygen. When acetate is broken down in methanogenesis, the result is the release of methane into the surrounding environment.

Methanogenesis, the scientific term for methane production, occurs primarily in anaerobic conditions because of the lack of availability of other oxidants. In these conditions, microscopic organisms called archaea use acetate and hydrogen to break down essential resources in a process called fermentation.

Acetoclastic methanogenesis: Certain archaea cleave acetate produced during anaerobic fermentation to yield methane and carbon dioxide.

$$H_3C\text{-}COOH \rightarrow CH_4 + CO_2$$

Hydrogenotrophic methanogenesis: Archaea oxidize hydrogen with carbon dioxide to yield methane and water.

$$4H_2 + CO_2 \rightarrow CH_4 + 2H_2O$$

While acetoclastic methanogenesis and hydrogenotrophic methanogenesis are the two major source reactions for atmospheric methane, other minor biological methane source reactions also occur. For example, it has been discovered that leaf surface wax exposed to UV radiation in the presence of oxygen is an aerobic source of methane.

Wetlands

Wetlands account for approximately 20 percent of atmospheric methane through emissions from soils and plants. Wetlands counteract the sinking action that normally occurs with soil because of the high water table. The level of the water table represents the boundary be-tween anaerobic methane production and aerobic methane consumption. When the water table is low, the methane generated within the wetland soil has to come up through the soil and get past a deeper layer of methanotrophic bacteria, thereby reducing emission. Meth-ane transport by vascular plants can bypass this aerobic layer, thus increasing emission.

Animals

Ruminant animals, particularly cows and sheep, contain bacteria in their gastrointes-tinal systems that help to break down plant material. Some of these microorganisms use the acetate from the plant material to produce methane, and because these bacteria live in the stomachs and intestines of ruminants, whenever the animal "burps" or def-ecates, it emits methane as well. Based upon a study in the Snowy Mountains region, the amount of methane emitted by one cow is equivalent to the amount of methane that around 3.4 hectares of methanotrophic bacteria can consume.

Termites also contain methanogenic microorganisms in their gut. However, some of these microorganisms are so unique that they live nowhere else in the world except in the third gut of termites. These microorganisms also break down biotic components to produce ethanol, as well as methane byproduct. However, unlike ruminants who lose 20 percent of the energy from the plants they eat, termites only lose 2 percent of their energy in the process. Thus comparatively, termites do not have to eat as much food as rumi-nants to obtain the same amount of energy, and give off proportionally less methane.

Plants

Living plants (e.g. forests) have recently been identified as a potentially important source of methane, possibly being responsible for approximately 10 to 30 percent of atmospheric methane. A 2006 paper calculated emissions of 62–236 Tg a^{-1}, and "this newly identified source may have important implications". However the authors stress "our findings are preliminary with regard to the methane emission strength".

These findings have been called into question in a 2007 paper which found "there is no evidence for substantial aerobic methane emission by terrestrial plants, maximally 0.3% of the previously published values".

While the details of plant methane emissions have yet to be confirmed, plants as a sig-nificant methane source would help fill in the gaps of previous global methane budgets as well as explain large plumes of methane that have been observed over the tropics.

In wetlands, where the rate of methane production is high, plants help methane trav-el into the atmosphere—acting like inverted lightning rods as they direct the gas up through the soil and into the air. They are also suspected to produce methane them-selves, but because the plants would have to use aerobic conditions to produce meth-ane, the process itself is still unidentified.

Methane Gas from Methane Clathrates

At high pressures, such as are found on the bottom of the ocean, methane forms a solid clathrate with water, known as methane hydrate. An unknown, but possibly very large quantity of methane is trapped in this form in ocean sediments. The release of large volumes of methane gas from such sediments into the atmosphere has been suggested as a possible cause for rapid global warming events in the Earth's distant past, such as the Paleocene–Eocene Thermal Maximum of 55 million years ago, and the Great Dying.

Theories suggest that should global warming cause them to heat up sufficiently, all of this methane gas could again be released into the atmosphere. Since methane gas is twenty-five times stronger (for a given weight, averaged over 100 years) than CO_2 as a greenhouse gas; this would immensely magnify the greenhouse effect. However, most of this reservoir of hydrates appears isolated from changes to the surface climate, so any such release is likely to happen over geological timescales of a millennium or more.

Permafrost

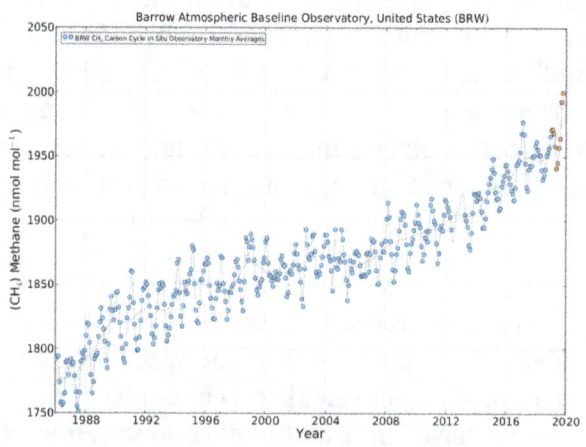

Arctic methane concentrations up to October 2019.

Methane that gets frozen in permafrost – land that is frozen for several years at a time – is slowly released from bogs as the permafrost melts. With rising global temperatures, the amount of permafrost melting and releasing methane continues to increase.

Although records of permafrost are limited, recent years have seen record thawing of permafrost in Alaska and Siberia. Measurements during 2006 in Siberia show that the methane released is five times greater than previously estimated. Melting yedoma, a type of permafrost, is a significant source of atmospheric methane (about 4 Tg of CH_4 per year).

The Woods Hole Research Center, citing two 2015 studies on permafrost carbon says there may be a self-reinforcing tipping point where an estimated equivalent of 205 gigatons of carbon dioxide in the form of methane could cause up to 0.5 °C (up to 0.9 °F) warming by the end of the century, which would trigger more warming. Permafrost contains almost twice as much carbon as is present in the atmosphere. The Intergovernmental Panel on Climate Change does not adequately account for arctic methane in permafrost.

Anthropogenic Sources of Atmospheric Methane

Slightly over half of the total emission is due to human activity. Since the Industrial Revolution humans have had a major impact on concentrations of atmospheric methane, increasing atmospheric concentrations roughly 250%.

Ecological Conversion

Conversion of forests and natural environments into agricultural plots increases the amount of nitrogen in the soil, which inhibits methane oxidation, weakening the ability of the methanotrophic bacteria in the soil to act as sinks. Additionally, by changing the level of the water table, humans can directly affect the soil's ability to act as a source or sink. The relationship between water table levels and methane emission is explained in the wetlands section of natural sources.

Farm Animals

A 2006 UN FAO report reported that livestock generate more greenhouse gases as measured in CO_2 equivalents than the entire transportation sector. Livestock accounts for 9 percent of anthropogenic CO_2, 65 percent of anthropogenic nitrous oxide and 37 percent of anthropogenic methane. A senior UN official and co-author of the report, Henning Steinfeld, said "Livestock are one of the most significant contributors to today's most serious environmental problems."

Recent NASA research has confirmed the vital role of enteric fermentation in livestock on global warming. "We understand that other greenhouse gases apart from carbon dioxide are important for climate change today," said Gavin Schmidt. Other recent peer reviewed NASA research published in the journal Science has also indicated that the contribution of methane to global warming has been underestimated.

Nicholas Stern, has stated "people will need to turn vegetarian if the world is to conquer climate change". President of the National Academy of Sciences Ralph Cicerone (an atmospheric scientist), has indicated the contribution of methane by livestock flatulence and eructation to global warming is a "serious topic." Cicerone states "Methane is the second-most-important greenhouse gas in the atmosphere now. The population of beef cattle and dairy cattle has grown so much that methane from cows now is big. This is not a trivial issue."

Approximately 5% of the methane is released via the flatus, whereas the other 95% is released via eructation. Vaccines are under development to reduce the amount introduced through eructation.

Rice Agriculture

Due to a continuously growing world population, rice agriculture has become one of the most significant anthropogenic sources of methane. With warm weather and water-logged soil, rice paddies act like wetlands, but are generated by humans for the purpose of food production. Due to the swamp-like environment of rice fields, these paddies yield 50–100 million metric tons of methane emission each year. This means that rice agriculture is responsible for approximately 15 to 20 percent of anthropogenic methane emissions. An article written by William F. Ruddiman explores the possibility that methane emissions began to rise as a result of anthropogenic activity 5000 years ago when ancient cultures started to settle and use agriculture, rice irrigation in particular, as a primary food source.

Landfills

Due to the large collections of organic matter and availability of anaerobic conditions, landfills are the third largest source of atmospheric methane in the United States, accounting for roughly 18.2% of methane emissions globally in 2014. When waste is first added to a landfill, oxygen is abundant and thus undergoes aerobic decomposition; during which time very little methane is produced. However, generally within a year oxygen levels are depleted and anaerobic conditions dominate the landfill allowing methanogens to takeover the decomposition process. These methanogens emit methane into the atmosphere and even after the landfill is closed, the mass amount of decaying matter allows the methanogens to continue producing methane for years.

Waste Water Treatment

Waste water treatment facilities act to remove organic matter, solids, pathogens, and chemical hazards as a result of human contamination. Methane emission in waste treatment facilities occurs as a result of anaerobic treatments of organic compounds and anaerobic biodegradation of sludge.

Biomass Burning

Incomplete burning of both living and dead organic matter results in the emission of methane. While natural wildfires can contribute to methane emissions, the bulk majority of biomass burning occurs as a result of humans – including everything from accidental burnings by civilians to deliberate burnings used to clear out land to biomass burnings occurring as a result of destroying waste.

Oil and Natural Gas Supply Chain

Methane is a primary component of natural gas, and thus during the production, processing, storage, transmission, and distribution of natural gas, a significant amount of methane is lost into the atmosphere.

According to the EPA Inventory of U.S Greenhouse Gas Emissions and Sinks: 1990–2015 report, 2015 methane emissions from natural gas and petroleum systems totaled 8.1 Tg per year in the United States. Individually, the EPA estimates that the natural gas system emitted 6.5 Tg per year of methane while petroleum systems emitted 1.6 Tg per year of methane. Methane emissions occur in all sectors of the natural gas industry, from drilling and production, through gathering and processing and transmission, to distribution. These emissions occur through normal operation, routine maintenance, fugitive leaks, system upsets, and venting of equipment. In the oil industry, some underground crude contains natural gas that is entrained in the oil at high reservoir pressures. When oil is removed from the reservoir, associated gas is produced.

However, a review of methane emissions studies reveals that the EPA Inventory of Greenhouse Gas Emissions and Sinks: 1990–2015 report likely significantly underestimated 2015 methane emissions from the oil and natural gas supply chain. The review concluded that in 2015 the oil and natural gas supply chain emitted 13 Tg per year of methane, which is about 60% more than the EPA report for the same time period. The authors write that the most likely cause for the discrepancy is an under sampling by the EPA of so-called "abnormal operating conditions", during which large quantities of methane can be emitted.

Methane Slip from Gas Engines

The use of natural gas and biogas in ICE (Internal combustion engine) for such applications as electricity production/cogeneration/CHP (Combined Heat and Power) and heavy vehicles or marine vessels such as LNG carriers using the boil off gas for propulsion, emits a certain percentage of UHC, unburned hydrocarbon of which 85% is methane. The climate issues of using gas to fuel ICE may offset or even cancel out the advantages of less CO_2 and particle emissions is described in this 2016 EU Issue Paper on methane slip from marine engines: "Emissions of unburnt methane (known as the 'methane slip') were around 7g per kg LNG at higher engine loads, rising to 23–36g at lower loads. This increase could be due to slow combustion at lower temperatures, which allows small quantities of gas to avoid the combustion process". Road vehicles run more on low load than marine engines causing relatively higher methane slip.

Coal Mining

In 2014 NASA researchers reported the discovery of a 2,500 square miles (6,500 km²) methane cloud floating over the Four Corners region of the south-west United States. The discovery was based on data from the European Space Agency's Scanning Imaging Absorption Spectrometer for Atmospheric Chartography instrument from 2002 to 2012.

The report concluded that "the source is likely from established gas, coal, and coalbed methane mining and processing." The region emitted 590,000 metric tons of methane every year between 2002 and 2012—almost 3.5 times the widely used estimates in the European Union's Emissions Database for Global Atmospheric Research.

Removal Processes

Any process that consumes methane from the atmosphere can be considered a "sink" of atmospheric methane. The most prominent of these processes occur as a result of methane either being destroyed in the atmosphere or broken down in soil. Humans have yet to act as any significant sink of atmospheric methane.

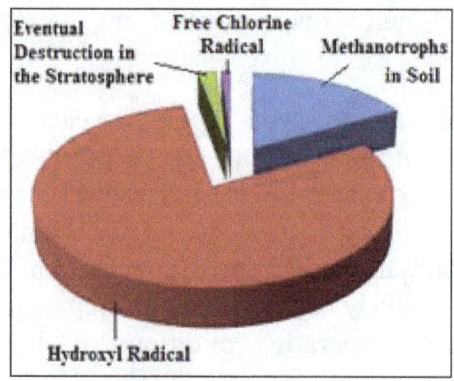

A pie chart demonstrating the relative effects of various sinks of atmospheric methane.

Reaction with the hydroxyl radical: The major removal mechanism of methane from the atmosphere involves radical chemistry; it reacts with the hydroxyl radical ($\cdot$OH) in the troposphere or stratosphere to create the $\cdot CH_3$ radical and water vapor. In addition to being the largest known sink for atmospheric methane, this reaction is one of the most important sources of water vapor in the upper atmosphere. Following the reaction of methane with the hydroxyl radical, two dominant pathways of methane oxidation exist: which leads to a net production of ozone, and which causes no net ozone change. For methane oxidation to take the pathway that leads to net ozone production, nitric oxide (NO) must be available to react with $CH_3O_2\cdot$. Otherwise, $CH_3O_2\cdot$ reacts with the hydroperoxyl radical ($HO_2\cdot$), and the oxidation takes the pathway with no net ozone change. Both oxidation pathways lead to a net production of formaldehyde and water vapor.

Net production of O_3

$$CH_4 + \cdot OH \rightarrow CH_3\cdot + H_2O$$

$$CH_3\cdot + O_2 + M \rightarrow CH_3O_2\cdot + M$$

$$CH_3O_2\cdot + NO \rightarrow NO_2 + CH_3O\cdot$$

$$CH_3O\cdot + O_2 \rightarrow HO_2\cdot + HCHO$$

$$HO_2\cdot + NO \rightarrow NO_2 + \cdot OH$$

$$(2x)\ NO_2 + hv \rightarrow O(^3P) + NO$$

$$(2x)\ O(^3P) + O_2 + M \rightarrow O_3 + M$$

$$[NET:\ CH_4 + 4O_2 \rightarrow HCHO + 2O_3 + H_2O]$$

No net change of O_3

$$CH_4 + \cdot OH \rightarrow CH_3\cdot + H_2O$$

$$CH_3\cdot + O_2 + M \rightarrow CH_3O_2\cdot + M$$

$$CH_3O_2\cdot + HO_2\cdot + M \rightarrow CH_3O_2H + O_2 + M$$

$$CH_3O_2H + hv \rightarrow CH_3O\cdot + \cdot OH$$

$$CH_3O\cdot + O_2 \rightarrow HO_2\cdot + HCHO$$

$$[NET:\ CH_4 + O_2 \rightarrow HCHO + H_2O]$$

Note that for the second reaction, there will be a net loss of radicals in the case where CH_3O_2H is lost to wet deposition before it can undergo photolysis such that: $CH_3O_2H + H_2O \rightarrow$ wet deposition. Also note that M represents a random molecule that facilitates energy transfer during the reaction.

This reaction in the troposphere gives a methane lifetime of 9.6 years. Two more minor sinks are soil sinks (160-year lifetime) and stratospheric loss by reaction with $\cdot OH$, $\cdot Cl$ and $\cdot O^1D$ in the stratosphere (120-year lifetime), giving a net lifetime of 8.4 years. Oxidation of methane is the main source of water vapor in the upper stratosphere (beginning at pressure levels around 10 kPa).

The methyl radical formed in the above reaction will, during normal daytime conditions in the troposphere, usually react with another hydroxyl radical to form formaldehyde. Note that this is not strictly oxidative pyrolysis as described previously. Formaldehyde can react again with a hydroxyl radical to form carbon dioxide and more water vapor. Sidechains in these reactions may interact with nitrogen compounds that will likely produce ozone, thus supplanting radicals required in the initial reaction.

Natural Sinks of Atmospheric Methane

Most natural sinks occur as a result of chemical reactions in the atmosphere as well as oxidation by methane consuming bacteria in Earth's soils.

Methanotrophs in Soils

Soils act as a major sink for atmospheric methane through the methanotrophic bacteria that reside within them. This occurs with two different types of bacteria. "High capacity-low affinity" methanotrophic bacteria grow in areas of high methane concentration, such as waterlogged soils in wetlands and other moist environments. And in areas of low methane concentration, "low capacity-high affinity" methanotrophic bacteria make use of the methane in the atmosphere to grow, rather than relying on methane in their immediate environment.

Forest soils act as good sinks for atmospheric methane because soils are optimally moist for methanotroph activity, and the movement of gases between soil and atmosphere (soil diffusivity) is high. With a lower water table, any methane in the soil has to make it past the methanotrophic bacteria before it can reach the atmosphere.

Wetland soils, however, are often sources of atmospheric methane rather than sinks because the water table is much higher, and the methane can be diffused fairly easily into the air without having to compete with the soil's methanotrophs.

Methanotrophic bacteria in soils: Methanotrophic bacteria that reside within soil use methane as a source of carbon in methane oxidation. Methane oxidation allows methanotrophic bacteria to use methane as a source of energy, reacting methane with oxygen and as a result producing carbon dioxide and water.

$$CH_4 + 2O_2 \rightarrow CO_2 + 2H_2O$$

Troposphere

The most effective sink of atmospheric methane is the hydroxyl radical in the troposphere, or the lowest portion of Earth's atmosphere. As methane rises into the air, it reacts with the hydroxyl radical to create water vapor and carbon dioxide. The lifespan of methane in the atmosphere was estimated at 9.6 years as of 2001; however, increasing emissions of methane over time reduce the concentration of the hydroxyl radical in the atmosphere. With less $OH°$ to react with, the lifespan of methane could also increase, resulting in greater concentrations of atmospheric methane.

Stratosphere

If it is not destroyed in the troposphere, methane will last approximately 120 years before it is eventually destroyed in Earth's next atmospheric layer: the stratosphere. Destruction in the stratosphere occurs the same way that it does in the troposphere: methane is oxidized to produce carbon dioxide and water vapor. Based on balloon-borne measurements since 1978, the abundance of stratospheric methane has increased by 13.4%±3.6% between 1978 and 2003.

Reaction with Free Chlorine

The reaction of methane and chlorine atoms acts as a primary sink of Cl atoms and is a primary source of hydrochloric acid (HCl) in the stratosphere.

$$CH_4 + Cl \rightarrow CH_3 + HCl$$

The HCl produced in this reaction leads to catalytic ozone destruction in the stratosphere.

Trends in Methane Levels over Time

Since the 1800s, atmospheric methane concentrations have increased annually at a rate of about 0.9%.

Global Trends in Methane Levels

Long term atmospheric measurements of methane by NOAA show that the build up of methane leveled off during the decade prior to 2006, after nearly tripling since pre-industrial times. Although scientists have yet to determine what caused this reduction in the rate of accumulation of atmospheric methane, it appears it could be due to reduced industrial emissions and drought in wetland areas.

Exceptions to this drop in growth rate occurred in 1991 and 1998 when growth rates increased suddenly to 14–15 nmol/mol per year for those years, nearly double the growth rates of the years before.

The 1991 spike is understood to be due to the volcanic eruption of Mt. Pinatubo in June of that year. Volcanoes affect atmospheric methane emissions when they erupt, releasing ash and sulfur dioxide into the air. As a result, photochemistry of plants is affected and the removal of methane via the tropospheric hydroxyl radical is reduced. However, growth rates quickly fell due to lower temperatures and global reduction in rainfall.

The cause of the 1998 spike is unresolved, but scientists are currently attributing it to a combination of increased wetland and rice field emissions as well as an increased amount of biomass burning. 1998 was also the warmest year since surface temperatures were first recorded, suggesting that anomalously high temperatures can induce elevated methane emission.

Data from 2007 suggested methane concentrations were beginning to rise again. This was confirmed in 2010 when a study showed methane levels were on the rise for the 3 years 2007 to 2009. After a decade of near-zero growth in methane levels, "globally averaged atmospheric methane increased by (approximately) 7 nmol/mol per year during 2007 and 2008. During the first half of 2009, globally averaged atmospheric CH_4 was (approximately) 7 nmol/mol greater than it was in 2008, suggesting that the increase will continue in 2009." From 2015 to 2019 sharp rises in levels of atmospheric methane have been recorded.

Methane emissions levels vary greatly depending on the local geography. For both natural and anthropogenic sources, higher temperatures and higher water levels result in the anaerobic environment that is necessary for methane production.

Natural Methane Cycles

Emissions of methane into the atmosphere are directly related to temperature and moisture. Thus, the natural environmental changes that occur during seasonal change act as a major control of methane emission. Additionally, even changes in temperature during the day can affect the amount of methane that is produced and consumed.

For example, plants that produce methane can emit as much as two to four times more methane during the day than during the night. This is directly related to the fact that plants tend to rely on solar energy to enact chemical processes.

Additionally, methane emissions are affected by the level of water sources. Seasonal flooding during the spring and summer naturally increases the amount of methane released into the air.

Changes due to Human Activity

Changes due to Pre-industrial Human Activity

The most clearly identified rise in atmospheric methane as a result of human activity occurred in the 1700s during the industrial revolution. As technology increased at a considerable rate, humans began to build factories and plants, burn fossil fuels for energy, and clear out forests and other vegetation for the purpose of building and agriculture. This growth continued to rise at a rate of almost 1 percent per year until around 1990 when growth rates dropped to almost zero.

A 2003 article from William F. Ruddiman, however, indicates that the anthropogenic change in methane may have started 5000 years prior to the industrial revolution. The methane insolation cycles of the ice core remained stable and predictable until 5000 years ago, most likely due to some anthropogenic effect. Ruddiman suggests that the transition of humans from hunter gatherers into agricultural farming was the first instance of humans affecting methane concentration in the atmosphere. Ruddiman's hypothesis is supported by the fact that early rice irrigation occurred approximately 5000 years ago—the same time the ice core cycles lost their predictability. Due to the inefficiency of humans first learning how to grow rice, extensive rice paddies would have been needed to feed even a small population. These, over-flooded and filled with weeds, would have resulted in huge methane emitting wetlands.

Changes due to Industrial Human Activity

Increases in methane levels due to modern human activities arise from a number of specific sources:

- Methane emissions from industrial activity.

- Methane emissions from extraction of oil and natural gas from underground reserves.

- Methane emissions from transportation via pipeline of oil and natural gas.

- Methane emissions from melting permafrost in Arctic regions, due to global warming which is caused by human use of fossil fuels.

Emissions due to Oil and Gas Extraction

Natural Gas Pipelines

One source of methane emissions has been identified as pipelines that transport natural gas; one example is pipelines from Russia to customers in Europe. Near Yamburg and Urengoy exist gas fields with a methane concentration of 97 percent. The gas obtained from these fields is taken and exported to Western and Central Europe through an extensive pipeline system known as the Trans-Siberian natural gas pipeline system. In accordance with the IPCC and other natural gas emissions control groups, measurements had to be taken throughout the pipeline to measure methane emissions from technological discharges and leaks at the pipeline fittings and vents. Although the majority of the natural gas leaks were carbon dioxide, a significant amount of methane was also being consistently released from the pipeline as a result of leaks and breakdowns. In 2001, natural gas emissions from the pipeline and natural gas transportation system accounted for 1 percent of the natural gas produced. Fortunately, between 2001 and 2005, this number reduced to 0.7 percent, and even the 2001 value is still significantly less than that of 1996.

General Industrial Causes

However, pipeline transportation is only one part of the problem. Howarth et al. have argued that:

> We believe the preponderance of evidence indicates shale gas has a larger GHG (green house gas) footprint than conventional gas, considered over any time scale. The GHG footprint of shale gas also exceeds that of oil or coal when considered at decadal time scales.

For subsequent works confirming these results see Howarth's "A bridge to nowhere: methane emissions and the greenhouse gas footprint of natural gas," "Methane emissions and climatic warming risk from hydraulic fracturing and shale gas development: implications for policy," and "Ideas and perspectives: is shale gas a major driver of recent increase in global atmospheric methane?" A 2013 study by Miller et al. indicates that current greenhouse gas reduction policies in the US are based on what appear to be significant underestimates of anthropogenic methane emissions. The authors state:

> We find greenhouse gas emissions from agriculture and fossil fuel extraction

and processing (i.e., oil and natural gas) are likely a factor of two or greater than cited in existing studies.

Release of Stored Arctic Methane due to Global Warming

Global warming due to fossil fuel emissions has caused Arctic methane release, i.e. the release of methane from seas and soils in permafrost regions of the Arctic. Although in the long term, this is a natural process, methane release is being exacerbated and accelerated by global warming. This results in negative effects, as methane is itself a powerful greenhouse gas.

The Arctic region is one of the many natural sources of the greenhouse gas methane. Global warming accelerates its release, due to both release of methane from existing stores, and from methanogenesis in rotting biomass. Large quantities of methane are stored in the Arctic in natural gas deposits, permafrost, and as undersea clathrates. Permafrost and clathrates degrade on warming, thus large releases of methane from these sources may arise as a result of global warming. Other sources of methane include submarine taliks, river transport, ice complex retreat, submarine permafrost and decaying gas hydrate deposits.

Atmospheric Impacts

The direct radiative greenhouse gas forcing effect has been estimated at 0.5 W/m^2.

Methane is a strong GHG with a global warming potential 84 times greater than CO_2 in a 20-year time frame. Methane is not as persistent a gas and tails off to about 28 times greater than CO_2 for a 100-year time frame.

In addition to the direct heating effect and the normal feedbacks, the methane breaks down to carbon dioxide and water. This water is often above the tropopause where little water usually reaches. Ramanathan notes that both water and ice clouds, when formed at cold lower stratospheric temperatures, are extremely efficient in enhancing the atmospheric greenhouse effect. He also notes that there is a distinct possibility that large increases in future methane may lead to a surface warming that increases nonlinearly with the methane concentration.

Ozone Layer

Methane also affects the degradation of the ozone layer, when methane is transformed into water in the stratosphere. This process is enhanced by global warming, because warmer air holds more water vapor than colder air, so the amount of water vapor in the atmosphere increases as it is warmed by the greenhouse effect. Climate models also indicate that greenhouse gases such as carbon dioxide and methane may enhance the transport of water into the stratosphere; though this is not fully understood.

Methane Management Techniques

In an effort to mitigate climate change, humans have started to develop alternative methods and medicines.

For example, in order to counteract the amount of methane that ruminants give off, a type of drug called monensin (marketed as rumensin™) has been developed. This drug is classified as an ionophore, which is an antibiotic that is naturally produced by a harmless bacteria strain. This drug not only improves feed efficiency but also reduces the amount of methane gas emitted from the animal and its manure.

In addition to medicine, specific manure management techniques have been developed to counteract emissions from livestock manure. Educational resources have begun to be provided for small farms. Management techniques include daily pickup and storage of manure in a completely closed off storage facility that will prevent runoff from making it into bodies of water. The manure can then be kept in storage until it is either reused for fertilizer or taken away and stored in an offsite compost. Nutrient levels of various animal manures are provided for optimal use as compost for gardens and agriculture.

In order to reduce effects on methane oxidation in soil, several steps can be taken. Controlling the usage of nitrogen enhancing fertilizer and reducing the amount of nitrogen pollution into the air can both lower inhibition of methane oxidation. Additionally, using drier growing conditions for crops such as rice and selecting strains of crops that produce more food per unit area can reduce the amount of land with ideal conditions for methanogenesis. Careful selection of areas of land conversion (for example, plowing down forests to create agricultural fields) can also reduce the destruction of major areas of methane oxidation.

To counteract methane emissions from landfills, on March 12, 1996, the EPA (Environmental Protection Agency) added the "Landfill Rule" to the Clean Air Act. This rule requires large landfills that have ever accepted municipal solid waste, have been used as of November 8, 1987, can hold at least 2.5 million metric tons of waste with a volume greater than 2.5 million cubic meters, and have nonmethane organic compound (NMOC) emissions of at least 50 metric tons per year to collect and combust emitted landfill gas. This set of requirements excludes 96% of the landfills in the USA. While the direct result of this is landfills reducing emission of non-methane compounds that form smog, the indirect result is reduction of methane emissions as well.

Furthermore, in an attempt to absorb the methane that is already being produced from landfills, experiments in which nutrients were added to the soil to allow methanotrophs to thrive have been conducted. These nutrient supplemented landfills have been shown to act as a small scale methane sink, allowing the abundance of methanotrophs to sponge the methane from the air to use as energy, effectively reducing the landfill's emissions.

To reduce emissions from the natural gas industries, the EPA developed the Natural Gas STAR Program, also known as Gas STAR.

Another program was also developed by the EPA to reduce emissions from coal mining. The Coalbed Methane Outreach Program (CMOP) helps and encourages the mining industry to find ways to use or sell methane that would otherwise be released from the coal mine into the atmosphere.

Methane Emissions Monitoring

A portable methane detector has been developed which, mounted in a vehicle, can detect excess levels of methane in the ambient atmosphere and differentiate between natural methane from rotting vegetation or manure and gas leaks. As of 2013 the technology was being deployed by Pacific Gas and Electric.

Measurement of Atmospheric Methane

Gas Chromatography

Methane is typically measured using gas chromatography. Gas chromatography is a type of chromatography used for separating or analyzing chemical compounds. It is less expensive in general, compared to more advanced methods, but it is more time and labor-intensive.

Spectroscopic Method

Spectroscopic methods are the preferred method for atmospheric gas measurements due to its sensitivity and precision. Also, spectroscopic methods are the only way of remotely sensing the atmospheric gases. Infrared spectroscopy covers a large spectrum of techniques, one of which detects gases based on absorption spectroscopy. There are various methods for spectroscopic methods, including Differential optical absorption spectroscopy, Laser-induced fluorescence, and Fourier Transform Infrared.

Cavity ring-down Spectroscopy

Cavity ring-down spectroscopy is most widely used IR absorption technique of detecting methane. It is a form of laser absorption spectroscopy which determines the mole fraction to the order of parts per trillion.

TETRAFLUOROMETHANE

Tetrafluoromethane, also known as carbon tetrafluoride or R-14, is the simplest fluorocarbon (CF_4). It has a very high bond strength due to the nature of the carbon–fluorine bond. It can also be classified as a haloalkane or halomethane. Tetrafluoromethane is a useful refrigerant but also a potent greenhouse gas.

Tetrafluoromethane is a potent greenhouse gas that contributes to the greenhouse effect. It is very stable, has an atmospheric lifetime of 50,000 years, and a high greenhouse warming potential of 6500 (which is given for the first 100 years thereof, CO_2 has a factor of 1); however, the low amount in the atmosphere restricts the overall radiative forcing effect.

Although structurally similar to chlorofluorocarbons (CFCs), tetrafluoromethane does not deplete the ozone layer. This is because the depletion is caused by the chlorine atoms in CFCs, which dissociate when struck by UV radiation. Carbon–fluorine bonds are stronger and less likely to dissociate. According to Guinness World Records Tetrafluoromethane is the most persistent greenhouse gas.

Main industrial emissions of tetrafluoromethane besides hexafluoroethane are produced during production of aluminium using Hall-Héroult process. CF_4 also is produced as product of the breakdown of more complex compounds such as halocarbons.

Bonding

Because of the multiple carbon–fluorine bonds, and the high electronegativity of fluorine, the carbon in tetrafluoromethane has a significant positive partial charge which strengthens and shortens the four carbon–fluorine bonds by providing additional ionic character. Carbon–fluorine bonds are the strongest single bonds in organic chemistry. Additionally, they strengthen as more carbon–fluorine bonds are added to the same carbon. In the one carbon organofluorine compounds represented by molecules of fluoromethane, difluoromethane, trifluoromethane, and tetrafluoromethane, the carbon–fluorine bonds are strongest in tetrafluoromethane. This effect is due to the increased coulombic attractions between the fluorine atoms and the carbon because the carbon has a positive partial charge of 0.76.

Preparation

Tetrafluoromethane is the product when any carbon compound, including carbon itself, is burned in an atmosphere of fluorine. With hydrocarbons, hydrogen fluoride is a coproduct. It was first reported in 1926. It can also be prepared by the fluorination of carbon dioxide, carbon monoxide or phosgene with sulfur tetrafluoride. Commercially it is manufactured by the reaction of hydrogen fluoride with dichlorodifluoromethane or chlorotrifluoromethane; it is also produced during the electrolysis of metal fluorides MF, MF_2 using a carbon electrode.

Although it can be made from myriad precursors and fluorine, elemental fluorine is expensive and difficult to handle. Consequently, CF_4 is prepared on an industrial scale using hydrogen fluoride:

$$CCl_2F_2 + 2\,HF \rightarrow CF_4 + 2\,HCl$$

Laboratory Synthesis

Tetrafluoromethane can be prepared in the laboratory by the reaction of silicon carbide with fluorine:

$$SiC + 4\,F_2 \rightarrow CF_4 + SiF_4$$

Reactions

Tetrafluoromethane, like other fluorocarbons, is very stable due to the strength of its carbon–fluorine bonds. The bonds in tetrafluoromethane have a bonding energy of 515 $kJmol^{-1}$. As a result, it is inert to acids and hydroxides. However, it reacts explosively with alkali metals. Thermal decomposition or combustion of CF_4 produces toxic gases (carbonyl fluoride and carbon monoxide) and in the presence of water will also yield hydrogen fluoride.

It is very slightly soluble in water (about 20 mg L^{-1}), but miscible with organic solvents.

Uses

Tetrafluoromethane is sometimes used as a low temperature refrigerant (R-14). It is used in electronics microfabrication alone or in combination with oxygen as a plasma etchant for silicon, silicon dioxide, and silicon nitride. It also has uses in neutron detectors.

Health Risks

Due to its density, tetrafluoromethane can displace air, creating an asphyxiation hazard in inadequately ventilated areas.

NITROGEN TRIFLUORIDE

Nitrogen trifluoride is the inorganic compound with the formula NF_3. This nitrogen-fluorine compound is a colorless, nonflammable gas with a slightly musty odor. It finds increasing use as an etchant in microelectronics. Nitrogen trifluoride is an extremely strong greenhouse gas.

NF_3 is a greenhouse gas, with a global warming potential (GWP) 17,200 times greater than that of CO_2 when compared over a 100-year period. Its GWP place it second only to SF_6 in the group of Kyoto-recognised greenhouse gases, and NF_3 was included in that grouping with effect from 2013 and the commencement of the second commitment period of the Kyoto Protocol. It has an estimated atmospheric lifetime of 740 years, although other work suggests a slightly shorter lifetime of 550 years (and a corresponding GWP of 16,800).

Although NF_3 has a high GWP, for a long time its radiative forcing in the Earth's atmosphere has been assumed to be small, spuriously presuming that only small quantities are released into the atmosphere. Industrial applications of NF_3 routinely break it down, while in the past previously used regulated compounds such as SF_6 and PFCs were often released. Research has questioned the previous assumptions. High-volume applications such as DRAM computer memory production, the manufacturing of flat panel displays and the large-scale production of thin-film solar cells use NF_3.

Since 1992, when less than 100 tons were produced, production has grown to an estimated 4000 tons in 2007 and is projected to increase significantly. World production of NF_3 is expected to reach 8000 tons a year by 2010. By far the world's largest producer of NF_3 is the US industrial gas and chemical company Air Products and Chemicals. An estimated 2% of produced NF_3 is released into the atmosphere. Robson projected that the maximum atmospheric concentration is less than 0.16 parts per trillion (ppt) by volume, which will provide less than $0.001\ Wm^{-2}$ of IR forcing. The mean global tropospheric concentration of NF_3 has risen from about 0.02 ppt (parts per trillion, dry air mole fraction) in 1980, to 0.86 ppt in 2011, with a rate of increase of $0.095\ ppt\ yr^{-1}$, or about 11% per year, and an interhemispheric gradient that is consistent with emissions occurring overwhelmingly in the Northern Hemisphere, as expected. This rise rate in 2011 corresponds to about 1200 metric tons/y NF_3 emissions globally, or about 10% of the NF_3 global production estimates. This is a significantly higher percentage than has been estimated by industry, and thus strengthens the case for inventorying NF_3 production and for regulating its emissions. One study co-authored by industry representatives suggests that the contribution of the NF_3 emissions to the overall greenhouse gas budget of thin-film Si-solar cell manufacturing is overestimated. Instead, the contribution of the nitrogen trifluoride to the CO_2-budget of thin film solar cell production is compensated already within a few months by the CO_2 saving potential of the PV technology.

The UNFCCC, within the context of the Kyoto Protocol, decided to include nitrogen trifluoride in the second Kyoto Protocol compliance period, which begins in 2012 and ends in either 2017 or 2020. Following suit, the WBCSD/WRI GHG Protocol is amending all of its standards (corporate, product and Scope 3) to also cover NF_3.

Synthesis and Reactivity

Nitrogen trifluoride is a rare example of a binary fluoride that can be prepared directly from the elements only at very uncommon conditions, such as electric discharge. After first attempting the synthesis in 1903, Otto Ruff prepared nitrogen trifluoride by the electrolysis of a molten mixture of ammonium fluoride and hydrogen fluoride. It proved to be far less reactive than the other nitrogen trihalides nitrogen trichloride, nitrogen tribromide and nitrogen triiodide, all of which are explosive. Alone among the nitrogen trihalides it has a negative enthalpy of formation. Today, it is prepared both by direct reaction of ammonia and fluorine and by a variation of Ruff's method. It is supplied in pressurized cylinders.

Reactions

NF_3 is slightly soluble in water without undergoing chemical reaction. It is nonbasic with a low dipole moment of 0.2340 D. By contrast, ammonia is basic and highly polar (1.47 D). This difference arises from the fluorine atoms acting as electron withdrawing groups, attracting essentially all of the lone pair electrons on the nitrogen atom. NF_3 is a potent yet sluggish oxidizer.

It oxidizes hydrogen chloride to chlorine:

$$2\ NF_3 + 6\ HCl \rightarrow 6\ HF + N_2 + 3\ Cl_2$$

It converts to tetrafluorohydrazine upon contact with metals, but only at high temperatures:

$$2\ NF_3 + Cu \rightarrow N_2F_4 + CuF_2$$

NF_3 reacts with fluorine and antimony pentafluoride to give the tetrafluoroammonium salt:

$$NF_3 + F_2 + SbF_5 \rightarrow NF^+_4 SbF^-_6$$

Applications

Nitrogen trifluoride is used in the plasma etching of silicon wafers. Today nitrogen trifluoride is predominantly employed in the cleaning of the PECVD chambers in the high-volume production of liquid-crystal displays and silicon-based thin-film solar cells. In these applications NF_3 is initially broken down in situ by a plasma. The resulting fluorine atoms are the active cleaning agents that attack the polysilicon, silicon nitride and silicon oxide. Nitrogen trifluoride can be used as well with tungsten silicide, and tungsten produced by CVD. NF_3 has been considered as an environmentally preferable substitute for sulfur hexafluoride or perfluorocarbons such as hexafluoroethane. The process utilization of the chemicals applied in plasma processes is typically below 20%. Therefore some of the PFCs and also some of the NF_3 always escape into the atmosphere. Modern gas abatement systems can decrease such emissions.

Elemental fluorine has been introduced as an environmentally friendly replacement for nitrogen trifluoride in the manufacture of flat-panel displays and thin-film solar cells.

Nitrogen trifluoride is also used in hydrogen fluoride and deuterium fluoride lasers, which are types of chemical lasers. It is preferred to fluorine gas due to its convenient handling properties, reflecting its considerable stability.

It is compatible with steel and Monel, as well as several plastics.

Safety

Skin contact with NF_3 is not hazardous, and it is a relatively minor irritant to mucous membranes and eyes. It is a pulmonary irritant with a toxicity considerably lower than nitrogen oxides, and overexposure via inhalation causes the conversion of hemoglobin in blood to methemoglobin, which can lead to the condition methemoglobinemia. The National Institute for Occupational Safety and Health (NIOSH) specifies that the concentration that is immediately dangerous to life or health (IDLH value) is 1,000 ppm.

NITROUS OXIDE

Nitrous oxide, or N_2O, is a greenhouse gas that affects the ozone layer and the earth's climate. Until now, experts believed that microbes in the soil were largely responsible for its formation. The earth's flora emits considerable amounts of nitrous oxide that contributes to the greenhouse gas effect. Unlike human-induced global warming, however, this process is part of a natural effect.

Until now, climate reports like those from the UN's IPCC did not include plants as a significant source of nitrous oxide in the global climate budget. Yet to accurately calculate the human contribution to the greenhouse gas effect, it is essential to identify and quantify all sources of greenhouse gases – including the natural ones. The current study shows that all the plants studied emit nitrous oxide and contribute significantly to total N_2O emissions. The researchers report that based on these studies, emissions from plants could make up roughly five to ten percent of nitrous oxide in the earth's atmosphere. "To truly understand the role of plants in the nitrous oxide cycle and to quantify it more precisely, further studies on representative types of plants, especially trees, are needed," stresses Prof. Dr. Katharina Lenhart. "This study was just a first step toward quantifying plant emissions of nitrous oxide and understanding the related biochemical processes.

To determine the amount of N_2O emissions, the researchers studied 34 different plants under controlled conditions in a closed laboratory. Among the plants were tobacco, corn, and lavender. To avoid contamination with bacteria-generated nitrous oxide, some of the experiments were conducted under sterile conditions. All the experiments took place in the dark so that the nitrous oxide emitted could be related to plant respiration. Like humans, plants release carbon dioxide (CO_2) when they breathe. The opposite and generally better known process of CO_2 absorption, however, occurs only in the presence of light during photosynthesis. "The N_2O and CO_2 ratio is correlated, so we were able to use the ample existing research on plant carbon dioxide emissions to calculate the amount of nitrous oxide released," explains Prof. Lenhart.

Isotope analyses were also carried out, because all nitrous-oxide-producing processes release a nitrous oxide molecule with a typical isotope fingerprint, including plants. "By measuring the composition of the isotopes, we were able to clearly demonstrate that most of the nitrous oxide is not released by bacteria in the soil, and that it differs from all the previously known sources," adds Prof. Dr. Frank Keppler. In the next phase, the researchers will verify their laboratory results in field studies and include other plant species in their investigations. They also want to explore which biochemical process contributes to the formation of nitrous oxide in plants and the role of the biosphere in nitrous oxide formation in geological history. One particularly interesting question is how increasing global temperatures affect the rate at which plants release nitrous oxide.

CHLORODIFLUOROMETHANE AND TETRAFLUOROETHANE

HCFC-22 (chlorodifluoromethane, $CHClF_2$), and HFC-134a (1,1,1,2-tetrafluoroethane, CH_2FCF_3) are major coolants used in domestic and commercial refrigeration and air conditioning. Because of their ubiquitous use, these three gases are the most abundant of the chlorofluorocarbon (CFC), the hydrochlorofluorocarbon (HCFC), and the hydrofluorocarbon (HFC) species categories, respectively, in Earth's atmosphere. Due to their role in depleting stratospheric ozone, production and consumption of all CFCs were scheduled to be phased out gradually by the Montreal Protocol, first from developed countries by 1996, followed by developing countries by 2010. HCFC-22 has an ozone-depleting potential (ODP) about 20 times less than that of CFC-12 and so partly became an interim replacement for CFC-12 beginning in the late 1980s (HCFC-22 has had other significant uses such as propellants and foam blowing beginning in the early 1970s). Production and emissions of HFC-134a began slightly later, in the early 1990s, as a preferred component of motor vehicle air conditioning systems to replace CFC-12. Both HCFC-22 and HFC-134a are potent greenhouse gases, with global warming potentials (GWPs) of 1,760 and 1,300 on a 100-y time scale. The production of HFC-134a is anticipated to continue, and its emissions will very likely increase, until transition is made to refrigerants with low ODPs and GWPs.

Country-based annual consumption and production magnitudes for the HCFCs and emission data for HFC-134a have been collected since the 1990s, by the United Nations

Environment Program (UNEP) and the United Nations Framework Convention on Climate Change (UNFCCC), respectively. National-scale emissions are difficult to accurately quantify from these consumption and production data. First, there is usually a delay time from production of these gases to actual atmospheric release. This delay, known as the bank effect, ranges from near zero to decades, depending on how the chemicals are used. Second, production and emission data from reporting countries are neither audited nor independently verified under the Montreal and the Kyoto Protocols. Finally, HFC emissions reported by the UNFCCC do not represent global totals because they do not currently include emissions from developing countries.

In contrast to these "bottom–up" estimates, a number of other studies have used observations of atmospheric concentrations, atmospheric transport models, and inversion methods to derive regional-scale emissions. These so-called "top–down" studies can improve the spatial and temporal distribution of bottom–up inventories and provide insights on the associated emission mechanisms that may be used as information for climate change mitigation. Most recent top–down studies of HCFC-22 and HFC-134a indicate lower US emissions and higher emissions in Europe, Australia, and Asia than in the existing inventories.

The seasonal variation in the emissions has been generally overlooked in the past. Refrigerant leakage studies suggest neither the gradual leaks (i.e., regular emissions) nor the immediate release (i.e., break of the air conditioning system) have significant seasonal dependences. In contrast, a few regional atmospheric studies suggest that emissions of HCFC-22 and HFC-134a may be seasonal due to weather-dependent patterns in refrigerator and air conditioner use. This uncertainty in seasonality indicates incomplete knowledge of the underlying mechanisms that lead to atmospheric release. Effective emission control strategies depend on understanding the processes that cause these refrigerant emissions.

CHLOROFLUOROCARBON

Chlorofluorocarbon (CFC) is an organic compound that contains carbon, chlorine, and fluorine, produced as a volatile derivative of methane and ethane. A common subclass is the hydrochlorofluorocarbons (HCFCs), which contain hydrogen, as well. Freon is DuPont's brand name for CFCs, HCFCs and related compounds. Other commercial names from around the world are Algofrene, Arcton, Asahiflon, Daiflon, Eskimo, FCC, Flon, Flugene, Forane, Fridohna, Frigen, Frigedohn, Genetron, Isceon, Isotron, Kaiser, Kaltron, Khladon, Ledon, Racon, and Ucon. The most common representative is dichlorodifluoromethane (R-12 or Freon-12).

Chlorofluorocarbons (CFCs) are a family of chemical compounds developed back in the 1930's as safe, non-toxic, non-flammable alternative to dangerous substances like ammonia for purposes of refrigeration and spray can propellants. Their usage grew enormously over the years. One of the elements that make up CFCs is chlorine. Very little chlorine exists naturally in the atmosphere. But it turns out that CFCs are an excellent way of introducing chlorine into the ozone layer. The ultraviolet radiation at this altitude breaks down CFCs, freeing the chlorine. Under the proper conditions, this chlorine has the potential to destroy large amounts of ozone. This has indeed been observed, especially over Antarctica. As a consequence, levels of genetically harmful ultraviolet radiation have increased.

Production of new stocks ceased in most countries as of 1994. However many countries still require aircraft to be fitted with halon fire suppression systems because no safe and completely satisfactory alternative has been discovered for this application. There are also a few other, highly specialized uses. These programs recycle halon through "halon banks" coordinated by the Halon Recycling Corporation to ensure that discharge to the atmosphere occurs only in a genuine emergency and to conserve remaining stocks.

Development of Alternatives for CFCs

Work on alternatives for chlorofluorocarbons in refrigerants began in the late 1970s after the first warnings of damage to stratospheric ozone were published. The hydro-chlorofluorocarbons (HCFCs) are less stable in the lower atmosphere, enabling them to break down before reaching the ozone layer. Nevertheless, a significant fraction of the HCFCs do break down in the stratosphere and they have contributed to more chlorine buildup there than originally predicted. Later alternatives lacking the chlorine, the hydrofluorocarbons (HFCs) have an even shorter lifetimes in the lower atmosphere. One of these compounds, HFC-134a, is now used in place of CFC-12 in automobile air conditioners. Hydrocarbon refrigerants (a propane/isobutane blend) are also used extensively in mobile air conditioning systems in Australia, the USA and many other countries, as they have excellent thermodynamic properties and perform particularly well in high ambient temperatures. One of the natural refrigerants (along with Ammonia and Carbon Dioxide), hydrocarbons have negligible environmental impacts and are also used worldwide in domestic and commercial refrigeration applications, and are s.

Applications and replacements for CFCs		
Application	Previously used CFC	Replacement
Refrigeration and air-conditioning	CFC-12 (CCl_2F_2); CFC-11(CCl_3F); CFC-13$(CClF_3)$; HCFC-22 $(CHClF_2)$; CFC-113 $(Cl_2FCCClF_2)$; CFC-114 $(CClF_2CClF_2)$; CFC-115 (CF_3CClF_2);	HFC-23 (CHF_3); HFC-134a (CF_3CFH_2); HFC-507 (a 1:1 azeotropic mixture of HFC 125 (CF_3CHF2) and HFC-143a (CF_3CH_3)); HFC 410 (a 1:1 azeotropic mixture of HFC-32 (CF_2H_2) and HFC-125 (CF_3CF_2H))
Propellants in medicinal aerosols	CFC-114 $(CClF_2CClF_2)$	HFC-134a (CF_3CFH_2); HFC-227ea (CF_3CHFCF_3)
Blowing agents for foams	CFC-11 (CCl_3F); CFC 113 $(Cl_2FC-CClF_2)$; HCFC-141b (CCl_2FCH_3)	HFC-245fa $(CF_3CH_2CHF_2)$; HFC-365 mfc $(CF_3CH_2CF_2CH_3)$
Solvents, degreasing agents, cleaning agents	CFC-11 (CCl_3F); CFC-113 (CCl_2FCClF_2)	None

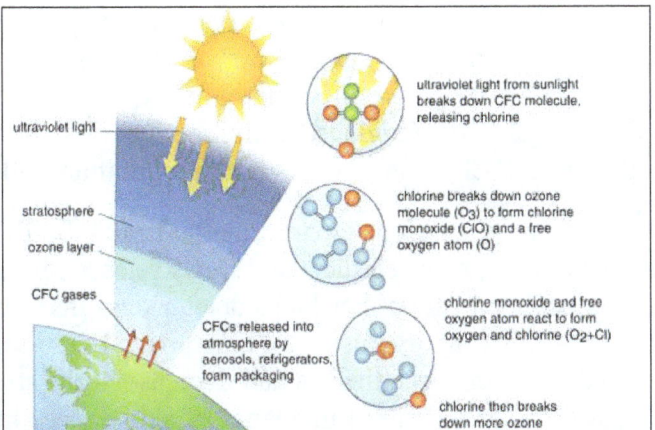

Refrigerators from the late 1800s until 1929 used the toxic gases, ammonia (NH_3), methyl chloride (CH_3Cl), and sulfur dioxide (SO_2), as refrigerants. Several fatal accidents occurred in the 1920s because of methyl chloride leakage from refrigerators. People started leaving their refrigerators in their backyards. A collaborative effort began between three American corporations, Frigidaire, General Motors and DuPont to search for a less dangerous method of refrigeration.

In 1928, Thomas Midgley, Jr. aided by Charles Franklin Kettering invented a "miracle compound" called Freon. Freon represents several different chlorofluorocarbons, or CFCs, which are used in commerce and industry. The CFCs are a group of aliphatic organic compounds containing the elements carbon and fluorine, and, in many cases, other halogens (especially chlorine) and hydrogen. Freons are colorless, odorless, non-flammable, noncorrosive gases or liquids.

Chlorofluorocarbons (CFCs) are highly stable compounds that were used as propellents in spray cans and in refrigeration units. They are several organic compounds composed of carbon, fluorine, chlorine, and hydrogen. CFCs are manufactured under the trade name Freon.

The invention of chlorofluorocarbons (CFCs) in the late 1920s and early 1930s stemmed from the call for safer alternatives to the sulfur dioxide and ammonia refrigerants used at the time, CFCs found wide application after World War II.

Chloroflourocarbons were first created in 1928 as non-toxic, non-flamable refrigerants, and were first produced commercially in the 1930's by DuPont. The first Chlorofluorocarbon was CFC-12, a single carbon with two chlorines and two Fluorines attached to it.

These halogenated hydrocarbons, notably trichlorofluoromethane (CFC-11, or F-11) and dichlorodifluoromethane (CFC-12, or F-12), have been used extensively as aerosol-spray propellants, refrigerants, solvents, and foam-blowing agents. They are well-suited for these and other applications because they are nontoxic and nonflammable and can be readily converted from a liquid to a gas and vice versa.

Chlorofluorocarbons or CFCs (also known as Freon) are non-toxic, non-flammable and non-carcinogenic. They contain fluorine atoms, carbon atoms and chlorine atoms. The 5 main CFCs include CFC-11 (trichlorofluoromethane - $CFCl_3$), CFC-12 (dichloro-di-fluoromethane - CF_2Cl_2), CFC-113 (trichloro-trifluoroethane - $C_2F_3Cl_3$), CFC-114 (di-chloro-tetrfluoroethane - $C_2F_4Cl_2$), and CFC-115 (chloropentafluoroethane - C_2F_5Cl).

CFCs have been found to pose a serious environmental threat. Studies undertaken by various scientists during the 1970s revealed that CFCs released into the atmosphere accumulate in the stratosphere, where they had a deleterious effect on the ozone layer. Stratospheric ozone shields living organisms on Earth from the harmful effects of the Sun's ultraviolet radiation; even a relatively small decrease in the stratospheric ozone concentration can result in an increased incidence of skin cancer in humans and in genetic damage in many organisms. In the stratosphere the CFC molecules break down by the action of solar ultraviolet radiation and release their constituent chlorine atoms. These then react with the ozone molecules, resulting in their removal.

CFCs have a lifetime in the atmosphere of about 20 to 100 years, and consequently one free chlorine atom from a CFC molecule can do a lot of damage, destroying ozone mol-ecules for a long time. Although emissions of CFCs around the developed world have largely ceased due to international control agreements, the damage to the stratospheric ozone layer will continue well into the 21st century.

In 1978, The Montreal Protocol was adopted as a framework for international coopera-tion regarding CFC control on the basis of the Vienna Convention for the Protection of the Ozone Layer.

FLUORINATED GASES

Fluorinated gases (F-gases) are man-made gases that can stay in the atmosphere for centuries and contribute to a global greenhouse effect. There are four types: hydroflu-orocarbons (HFCs), perfluorocarbons (PFCs), sulfur hexafluoride (SF_6) and nitrogen trifluoride (NF_3).

Types of F-gases

The most common F-gases are hydrofluorocarbons (HFCs), which contain hydro-gen, fluorine, and carbon. They are used in a multitude of applications including commercial refrigeration, industrial refrigeration, air-conditioning systems, heat pump equipment, and as blowing agents for foams, fire extinguishants, aerosol pro-pellants, and solvents.

Perfluorocarbons (PFCs) are the compounds consisting of fluorine and carbon. They are widely used in the electronics, cosmetics, and pharmaceutical industries, as well as in refrigeration when combined with other gases. PFCs were commonly used as fire extinguishers in the past and are still found in older fire protection systems. They are also a by-product of the aluminium smelting process.

Sulphur hexafluoride (SF_6) is used primarily as an insulation gas. It can be found in high-voltage switchgear and is used in the production of magnesium.

HFCs were developed in the 1990s to substitute substances such as chlorofluorocarbons (CFCs) and hydrochlorofluorocarbons (HCFCs). As these substances were found to deplete the ozone layer, the Montreal Protocol lays down provisions for them to be phased-out globally. PFCs and SF_6 were already in use prior to the Montreal Protocol.

Impact of F-gases

F-gases are ozone-friendly, very energy efficient, and safe for users and the public due to their low levels of toxicity and flammability. However, most F-gases have a relatively high global warming potential (GWP). If released, HFCs stay in the atmosphere for decades and both PFCs and SF_6 can stay in the atmosphere for millennia.

SULFUR HEXAFLUORIDE

Sulfur hexafluoride (SF_6) is an inorganic, colorless, odorless, non-flammable, non-toxic but extremely potent greenhouse gas, and an excellent electrical insulator. SF_6 has an octahedral geometry, consisting of six fluorine atoms attached to a central sulfur atom. It is a hypervalent molecule. Typical for a nonpolar gas, it is poorly soluble in water but quite soluble in nonpolar organic solvents. It is generally transported as a liquefied compressed gas. It has a density of 6.12 g/L at sea level conditions, considerably higher than the density of air (1.225 g/L).

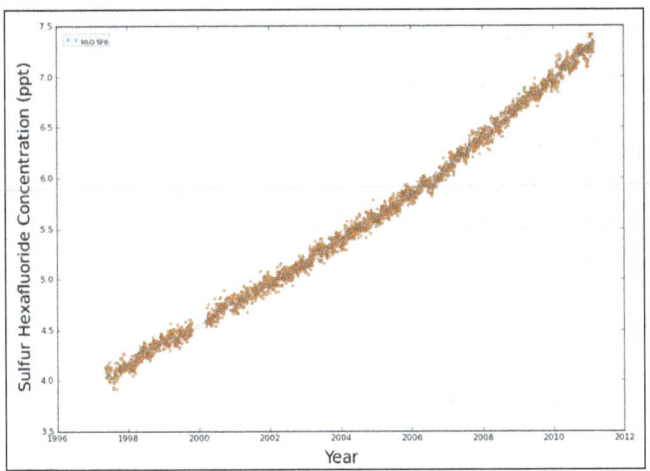

Mauna Loa sulfur hexafluoride timeseries.

According to the Intergovernmental Panel on Climate Change, SF$_6$ is the most potent greenhouse gas that it has evaluated, with a global warming potential of 23,900 times that of CO$_2$ when compared over a 100-year period. Sulfur hexafluoride is inert in the troposphere and stratosphere and is extremely long-lived, with an estimated atmospheric lifetime of 800–3,200 years.

Measurements of SF$_6$ show that its global average mixing ratio has increased by about 0.2 ppt (parts per trillion) per year to over 9 ppt as of February 2018. Average global SF$_6$ concentrations increased by about seven percent per year during the 1980s and 1990s, mostly as the result of its use in the magnesium production industry, and by electrical utilities and electronics manufacturers. Given the small amounts of SF$_6$ released compared to carbon dioxide, its overall contribution to global warming is estimated to be less than 0.2 percent.

In Europe, SF$_6$ falls under the F-Gas directive which ban or control its use for several applications. Since 1 January 2006, SF$_6$ is banned as a tracer gas and in all applications except high-voltage switchgear. It was reported in 2013 that a three-year effort by the United States Department of Energy to identify and fix leaks at its laboratories in the United States such as the Princeton Plasma Physics Laboratory, where the gas is used as a high voltage insulator, had been productive, cutting annual leaks by 35,000 pounds. This was done by comparing purchases with inventory, assuming the difference was leaked, then locating and fixing the leaks.

Synthesis and Reactions

SF$_6$ can be prepared from the elements through exposure of S$_8$ to F$_2$. This was also the method used by the discoverers Henri Moissan and Paul Lebeau in 1901. Some other sulfur fluorides are cogenerated, but these are removed by heating the mixture to disproportionate any S$_2$F$_{10}$ (which is highly toxic) and then scrubbing the product with NaOH to destroy remaining SF$_4$.

Alternatively, utilizing bromine, sulfur hexafluoride can be synthesized from SF$_4$ and CoF$_3$ at lower temperatures (e.g. 100 °C), as follows:

$$2CoF_3 + SF_4 + (Br_2) \rightarrow SF_6 + 2CoF_2 + (Br_2)$$

There is virtually no reaction chemistry for SF$_6$. A main contribution to the inertness of SF$_6$ is the steric hindrance of the sulfur atom, whereas its heavier group 16 counterparts, such as SeF$_6$ are more reactive than SF$_6$ as a result of less steric hindrance. It does not react with molten sodium below its boiling point, but reacts exothermically with lithium.

Applications

More than 10,000 tons of SF$_6$ are produced per year, most of which (over 8,000 tons) is used as a gaseous dielectric medium in the electrical industry. Other main uses include an inert gas for the casting of magnesium, and as an inert filling for insulated glazing windows.

Dielectric Medium

SF_6 is used in the electrical industry as a gaseous dielectric medium for high-voltage circuit breakers, switchgear, and other electrical equipment, often replacing oil filled circuit breakers (OCBs) that can contain harmful PCBs. SF_6 gas under pressure is used as an insulator in gas insulated switchgear (GIS) because it has a much higher dielectric strength than air or dry nitrogen. The high dielectric strength is a result of the gas's high electronegativity and density. This property makes it possible to significantly reduce the size of electrical gear. This makes GIS more suitable for certain purposes such as indoor placement, as opposed to air-insulated electrical gear, which takes up considerably more room. Gas-insulated electrical gear is also more resistant to the effects of pollution and climate, as well as being more reliable in long-term operation because of its controlled operating environment. Exposure to an arc chemically breaks down SF_6 though most of the decomposition products tend to quickly re-form SF_6, a process termed "self-healing". Arcing or corona can produce disulfur decafluoride (S_2F_{10}), a highly toxic gas, with toxicity similar to phosgene. S_2F_{10} was considered a potential chemical warfare agent in World War II because it does not produce lacrimation or skin irritation, thus providing little warning of exposure.

SF_6 is also commonly encountered as a high voltage dielectric in the high voltage supplies of particle accelerators, such as Van de Graaff generators and Pelletrons and high voltage transmission electron microscopes.

Alternatives to SF_6 as a dielectric gas include several fluoroketones.

Medical Use

SF_6 is used to provide a tamponade or plug of a retinal hole in retinal detachment repair operations in the form of a gas bubble. It is inert in the vitreous chamber and initially doubles its volume in 36 hours before being absorbed in the blood in 10–14 days.

SF_6 is used as a contrast agent for ultrasound imaging. Sulfur hexafluoride microbubbles are administered in solution through injection into a peripheral vein. These microbubbles enhance the visibility of blood vessels to ultrasound. This application has been used to examine the vascularity of tumours. It remains visible in the blood for 3 to 8 minutes, and is exhaled by the lungs.

Tracer Compound

Sulfur hexafluoride was the tracer gas used in the first roadway air dispersion model calibration; this research program was sponsored by the U.S. Environmental Protection Agency and conducted in Sunnyvale, California on U.S. Highway 101. Gaseous SF_6 is used as a tracer gas in short-term experiments of ventilation efficiency in buildings and indoor enclosures, and for determining infiltration rates. Two major factors recommend its use: its concentration can be measured with satisfactory accuracy at very low concentrations, and the Earth's atmosphere has a negligible concentration of SF_6.

Sulfur hexafluoride was used as a non-toxic test gas in an experiment at St. John's Wood tube station in London, United Kingdom on 25 March 2007. The gas was released throughout the station, and monitored as it drifted around. The purpose of the experiment, which had been announced earlier in March by the Secretary of State for Transport Douglas Alexander, was to investigate how toxic gas might spread throughout London Underground stations and buildings during a terrorist attack.

Sulfur hexafluoride is also routinely used as a tracer gas in laboratory fume hood containment testing. The gas is used in the final stage of ASHRAE 110 fume hood qualification. A plume of gas is generated inside of the fume hood and a battery of tests are performed while a gas analyzer arranged outside of the hood samples for SF_6 to verify the containment properties of the fume hood.

It has been used successfully as a tracer in oceanography to study diapycnal mixing and air-sea gas exchange.

Other Uses

- The United States Navy's Mark 50 torpedo closed Rankine-cycle propulsion system is powered by sulfur hexafluoride in an exothermic reaction with solid lithium.

- SF_6 plasma is also used in the semiconductor industry as an etchant. SF_6 breaks down in the plasma into sulfur and fluorine, the fluorine plasma performing the etching.

- The magnesium industry uses large amounts of SF_6 as inert gas to fill casting forms.

- Pressurizes waveguides in high-power microwave systems. The gas insulates the waveguide, preventing internal arcing.

- Has been used in electrostatic loudspeakers because of its high dielectric strength and high molecular weight.

- Was used to fill Nike Air bags in all of their shoes from 1992-2006.

- Feedstock for production of the chemical weapon disulfur decafluoride.

- For entertainment purposes, when breathed, SF_6 causes the voice to become significantly deeper, due to its density being so much higher than air, This is related to the more well-known effect of breathing low-density helium, which causes someone's voice to become much higher. Both of these effects should only be attempted with caution as these gases displace oxygen that the lungs are attempting to extract from the air. Sulfur hexafluoride is also mildly anesthetic.

- For science demonstrations/magic as "invisible water" since a light foil boat can be floated in a tank, as will an air filled balloon.

Physiological effects and Precautions

Like xenon, sulfur hexafluoride is a non-toxic gas, yet by displacing oxygen in the lungs, it also carries the risk of asphyxia if too much is inhaled. Being more dense than air, if a substantial quantity of gas is released it will settle in low-lying areas and present a significant risk of asphyxiation if the area is entered. This is particularly relevant to its use as an insulator in electrical equipment where workers may be in trenches or pits below equipment containing SF_6.

As with all gases, the density of SF_6 affects the resonance frequencies of the vocal tract, thus changing drastically the vocal sound qualities, or timbre, of those who inhale it. It does not affect the vibrations of the vocal folds. The density of sulfur hexafluoride is relatively high at room temperature and pressure due to the gas's large molar mass. Unlike helium, which has a molar mass of about 4 grams/mol and pitches the voice up, SF_6 has a molar mass of about 146 g/mol, and the speed of sound through the gas is about 134 m/s at room temperature, pitching the voice down. For comparison, the molar mass of air, which is about 80% nitrogen and 20% oxygen, is approximately 30 g/mol which leads to a speed of sound of 343 m/s.

Sulfur hexafluoride has an anesthetic potency slightly lower than nitrous oxide; Sulfur hexafluoride is classified as a mild anesthetic.

TRIFLUOROMETHYL SULFUR PENTAFLUORIDE

Trifluoromethyl sulfur pentafluoride (CF_3SF_5) is considered to be the most potent greenhouse gas present in the Earth's atmosphere. Its global warming potential is estimated to be 18 000 times that of carbon dioxide. CF_3SF_5 is resistant to photolysis and to reactions with common atmospheric and industrial ions, however it has been observed to react with positive ions in the ionosphere and undergo both electron attachment and protonation.

Recently, there has been a discovery of a previously unidentified greenhouse gas present in the atmosphere that is orders of magnitude more potent at inducing the greenhouse effect than is CO_2. This compound, trifluoromethyl sulfur pentafluoride (CF_3SF_5), is considered to be one of several super greenhouse gases, and because of its recent detection in the atmosphere, there is a relatively limited amount of literature available on CF_3SF_5.

Greenhouse gases (GHGs) have different capacities to contribute to global warming. This is due to differences in absorption and re-radiation at specific wavelengths in the infrared (IR) spectrum and from residence times in the atmosphere. The concept of the global warming potential (GWP) was introduced by the Intergovernmental Panel On Climate Change (IPCC) to express the relative effectiveness of a given amount of a specific GHG, over a specified period of time to induce a positive radiative forcing. This measure is based on the potency of carbon dioxide to induce global warming, given a GWP of 1, and then comparing all other GHGs to CO_2. Some common GHGs with their respective GWPs include methane (GWP=21) and nitrogen dioxide (GWP=310). Gases that have GWPs that are four to five times the magnitude of carbon dioxide are appropriately referred to as super greenhouse gases.

CF_3SF_5 is considered to be one of the several super greenhouse gases. On a per molecule basis, it is considered to be the most potent greenhouse gas present in Earth's atmosphere. The most effective greenhouse gases absorb radiation in the infrared window, containing the wavelengths from approximately 700 to 1300 cm^{-1}. Due to CF_3SF_5's strong absorptive power in this atmospheric window (60% of the wavelengths that are absorbed by CF_3SF_5 are contained within the 800-1300 cm^{-1} range), CF_3SF_5 is an extremely efficient inducer of the greenhouse effect. Furthermore, it has been calculated that CF_3SF_5 has a radiative force of 0.57 watt per square meter per part per billion [W m^{-2} ppb$_{-1}$]. Comparatively, some well known CFCs and HFCs have a radiative forcing per ppb ranging from 0.02 to 0.26 W m^{-2} ppb$_{-1}$. This data, provided in table, shows that CF_3SF_5 can be up to 25 to 30 times more potent at inducing the greenhouse effect than other chlorine and fluorine containing compounds that have been of interest in the past. By assuming that CF_3SF_5 has a similar atmospheric lifetime of SF_6, the 100-year global warming potential is estimated to be 18 000 times that of CO_2.

Chemistry in the Atmosphere

Several investigations have been conducted to determine how CF_3SF_5 is destroyed in the atmosphere and thereby provide an estimate of its lifetime. These mechanisms include photolysis, reactions with several atmospheric and industrial compounds, electron attachment, and protonation reactions. Table provides a summary of these data.

Table: Radiative Forcing of Selected Fluorinated Compounds.

Compound	Radiative Forcing (W m^{-2} ppb^{-1})
SF_6	0.64
CF_3SF_5	0.57
CFC 11 ($CFCl_3$)	0.29
HFC 134a (CH_2FCF_3)	0.17
HFC 152a (CH_3CHF_2)	0.11

Oxidation

The troposphere is an important area for chemical reactions to occur, as it contains

85% of total atmospheric mass. Oxidation is a primary route by which many trace gases are removed from the atmosphere, with the most abundant oxidizers being O_2 and O_3. However these molecules are generally uncreative towards non-radicals. Organic compounds are oxidized in the atmosphere primarily through reactions with OH and O_3 (also to some extent NO_3). Oxidation with hydroxyl radicals tends to form stable water molecules via abstraction of a hydrogen atom in alkanes or through double bond attack in alkenes forming alcohols. Ozone also attacks the double bond in alkenes. CF_3SF_5 contains neither the hydrogen atoms nor the double bonds that are the targets of OH and O_3 and therefore are unreactive to these oxidizers. No experimental data could be found, however oxidation modeling has been conducted with SF_6, which is molecularly similar to CF_3SF_5. It was calculated that SF_6 would have an atmospheric lifetime of 60000 years when removed exclusively by OH and O_3, thereby providing evidence that the oxidation of CF_3SF_5 is not likely to occur.

Table: Summary of Observed/Investigated Reactions with CF_3SF_5.

Type	Reactants	Products	Reaction Rate ($/10^{-9}cm^3 s^{-1}$)	Usual Atmospheric Region of Occurrence
Oxidation	OH, O_2, O_3	No reaction	---	Troposphere and Stratosphere
Photolysis	UV Photons	No reaction	---	Stratosphere
Anions	O_2^-	SF_5^-, CF_3, and O_2	0.36-0.55	Ionosphere (D region)
	CO_3^- and NO_3^-	No reaction	<0.006	Ionosphere (D region)
Cations	O_2^+	CF_3^+ and SF_3^+	0.01	Ionosphere (E region)
	NO^+	No reaction	---	Ionosphere (E region)
	N^+	CF_3^+ and SF_3^+	2.2	Ionosphere (F region)
	O^+	CF_3^+ and SF_3^+	1.9	Ionosphere (F region)
Electron Attachment	Electrons	SF_5^- and CF_3	77	Ionosphere
Protonation	Protons	CF_3, SF_4, HF, SF_3^+, and CF_3^+	---	---

Photolysis

Photolysis refers to reactions in which chemicals are altered by exposure to light, though it is usually UV photons that contain enough energy to break down molecules. Photolytic reactions play important roles in both the attenuation and activation of chemicals in the atmosphere. Gerstell et al. 2001 showed that the CF_3-SF_5 bond is able to withstand UV photo disassociation in the stratosphere. They hypothesized that because CF_3SF_5 contains no double bonds or hydrogen atoms, photolysis should be considered the primary destructive pathway for CF_3SF_5 in the atmosphere. If this is the case, CF_3SF_5 is expected to have a lifetime of approximately 4050 years in the atmosphere. No solar radiation of wavelengths shorter than 290 nm reaches the troposphere; therefore it is not likely to be destroyed in this region via photolysis as well.

Reactions with Ions

Charged particles are generally not found with any significant concentration in the troposphere and stratosphere and therefore would play a minor, if any role in the degradation of CF_3SF_5 in these areas. The charged particle concentration near ground level is on the order of 103 particles per cubic centimeter, however it is subject to a miniscule mixing ratio of approximately 1 part in 1016. This indicates that reactions with charged particles in the lower atmosphere will not likely destroy CF_3SF_5 molecules. Ions and electrons are most abundant at altitudes above 60 km, in the mesosphere and above because a majority of the Sun's ionizing radiation is absorbed at this altitude and above. Within the upper mesosphere and lower thermosphere, is the ionosphere, a region where ions are produced by photoionization. Chemistry in the ionosphere may be of lesser importance compared to actions occurring lower in the atmosphere; however, charged particle reactions can still represent a natural sink for CF_3SF_5 given that it is resistant to oxidation and photolysis.

The ionosphere is divided into several discrete sections and for the purposes of this discussion only the D, E, and F regions are of importance. The D region is located approximately 60 to 100 km in the atmosphere, contains some of the more complex molecular ions, and is also the only location where negative ions are observed in the ionosphere. The primary negative ion is O_2^-, however HCO_3^-, CO_3^-, CO_4^-, NO_3^-, and Cl^- are also present at various heights in the ionosphere. CF_3SF_5 has been experimentally shown to react with O_2^-, yielding SF_5^-, CF_3, and O_2. However, with measured rate constants ranging from 0.36 to 0.55 10- 9 cm3 /sec, it appears that O_2^- does not contribute to the atmospheric chemistry of CF_3SF_5. Arnold et al. 2003 observed no reaction between CF_3SF_5 and CO_3^- and NO_3^-.

The dominant ions located in the E region (100-150 km) are NO^+ and O_2^+, but the E region also contains a relatively large concentration of molecular ions as well. The dominant reaction mechanism between CF_3SF_5 and ions is most likely dissociative charge transfer, whereby CF_3SF_5 transfers an electron to the charged particle. Atturbury et al. 2001 showed that CF_3SF_5 can react with O_2^+ to form primarily CF_3^+ and SF_3^+, but determined that the rate coefficient for this reaction was 0.01 x 10-9 cm³/sec, indicating that this reaction does not significantly destroy CF_3SF_5 molecules in the atmosphere. Atturbury et al. 2001 also showed that CF_3SF_5 is unreactive with NO^+. Reactions with $N2^+$, CO^+, H_2O, CO_2^+, N_2O^+ were investigated, in which varying degrees of reactivity was observed. Although the reaction rate constants were all equal or greater than 1.1 x 10^{-9}cm³/sec, these species are known to exist in the upper atmosphere at very low concentrations and therefore is not likely that these ions will contribute significantly to the overall decomposition of CF_3SF_5.

Dominant ions in the F region (>150 km) include O^+ and N^+, while H^+ and He^+ are most abundant at the highest altitudes (i.e. >1000 km). The makeup of ions in the F region can be described as being comprised of primarily atomic ions. Total charged particle (positive and negative ions and electrons) concentrations are only approximately a few hundred particles per cubic centimeter at 300 km in the F region. It has been shown that CF_3SF_5 will disassociate in reactions with N^+, O^+, Ne^+, Ar^+, and F^+ under experimental conditions. N^+ and O^+ are the dominant ions in the F region of the ionosphere, however because of the relatively low atmospheric mass that exists in this region, it is unlikely that CF_3SF_5 will undergo significant disassociation. Ne^+, Ar^+, and F^+ are considered trace ions in the ionosphere and are expected to have an even more subdued effect on CF_3SF_5 than N^+ and O^+.

Electron Attachment and Protonation

Electron attachment reactions involve the attack of a free electron to a parent molecule forming an energetic and unstable intermediate ion that will then fragment. Kennedy and Mayhew 2001 present observations that CF_3SF_5 can be destroyed by fast dissociative electron attachment, yielding SF_5^- and CF_3 with a reaction rate constant of 7.7 x 10-8 cm^3/sec. Electron concentrations are greatest in the upper atmosphere; electrons range from ~104 to 106 particles per cm^3 at 100-250 km, the E and F regions of the ionosphere. Given the relatively large reaction rate constant, electron attachment reactions play an important role in the decomposition of CF_3SF_5 in the upper atmosphere.

Protonation refers to the addition of a proton to a parent molecule. As in electron attachment reactions, the formation of an unstable intermediary is formed and is subsequently dissociated in protonation reactions. The resulting products of this reaction are CF_3, SF_4 and HF, including the formation of the following ions: SF_3^+ and CF_3^+. Pepi et al. 2005 investigated these reactions with several proton sources, and of those studied, CO_2H+, N_2OH^+, and COH^+ appear to be most potent at reacting with CF_3SF_5. As in many of the observed ion reactions, these molecules are not dominant in either the stratosphere or in the ionosphere, while H+ are only present at very high altitudes (>1000 km).

Atmospheric Lifetime

The only significant loss process for CF_3SF_5 in the atmosphere appears to be electron attachment reactions that can occur in the ionosphere. As discussed previously, it is estimated that CF_3SF_5 will have a lifetime of 60 000 years if solely destroyed by oxidation mechanisms in the stratosphere or 4050 years if degraded exclusively by photolysis. The lifetime of perfluorocompounds (fully fluorinated compounds) has been previously modeled. These models indicate that for substances that are primarily inert, but subject to electron attachment, such as CF_3SF_5 and also SF_6, atmospheric lifetime is estimated to range from 800 to 1000 years. The electron attachment reaction rate for SF_6 in the atmosphere has been calculated to be approximately 310 x 10^{-9} cm^3/sec, representing a rate that is close to the theoretical maximum. Given that the electron attachment rate of CF_3SF_5 is less, though relatively similar to that of SF_6 and that CF_3SF_5 is resistant to all other degradation reactions, it is estimated that the atmospheric lifetime of CF_3SF_5 is on the order of 1000 years. For comparison, the estimated lifetime of CO_2, CH_4, and N_2O are 50 to 100, 12, and 120 years, respectively.

The lifetime of CF_3SF_5 is 10 to 100 times longer than these greenhouse gases that are causing significant problems today.

Current Rate of Increase in the Atmosphere

The discovery of CF_3SF_5 as a component of the atmospheric gas mixture was very much an inadvertent one. Sturges et al., 2000 reported that when analyzing for SF_6 in stratospheric gas samples, they noted a previously unidentified peak that eluded shortly after SF_6 in their gas chromatographic analyses. Subsequent investigations determined that these peaks corresponded to the presence of CF_3SF_5.

After the identification of the presence CF_3SF_5 in stratospheric gas samples, efforts began to determine the concentration of this compound in the atmosphere. In January of 1999, air samples were pumped out of deep consolidated snow (firn) at Dome Concordia in eastern Antarctica. These samples showed that the atmospheric concentration of CF_3SF_5 in the early 1960s were essentially zero; with measurable levels starting in the mid 1960s. Sturges et al. 2000 interpreted these data as emissions of CF_3SF_5 to have begun in the late 1950s. After initial emissions began, a steady increase in concentration can be observed into the 1990s. Current levels of CF_3SF_5 is approximately 0.12 to 0.18 parts per trillion (ppt), increasing approximately six percent per year based on analyses conducted on Antarctic firn. Sturges et al. reported a +/- 10 percent error to this predicted rate of increase.

Current concentrations of CF_3SF_5 in the atmosphere do not contribute to overall radiative forcing, however this may change as observed trends continue. For comparison, atmospheric concentrations of CH_4 and N_2O are 1750 parts per billion (ppb) and 310 ppb, respectively. It is estimated that anthropogenic releases of CH_4 must be reduced by 8% to current stabilize atmospheric levels, while releases of N_2O must be reduced by 50%. Current concentration of atmospheric CO_2 is approximately 360 parts per million (ppm). Carbon dioxide, methane, and nitrogen dioxide, have contributed to 64%, 19%, and 6%, respectively, to global warming since the beginning of the Industrial Era. Given that the atmospheric concentration of CF_3SF_5 is still very low, we have the unique opportunity to identify and mitigate sources of this molecule prior to it contributing to climate changes.

Possible Anthropogenic Sources

There are several possible uses of CF_3SF_5 in industry that may be contributing to the increased concentrations observed in the atmosphere. The estimate of zero concentration of CF_3SF_5 in the atmosphere prior to the 1960s indicates that the source of this molecule is exclusively anthropogenic. Santoro 2000 claims that he is aware of one definitive source of CF_3SF_5, a by-product of the manufacture of fluorochemicals. It has been presented that the source of CF_3SF_5 in the atmosphere may originate from reactions of SF_6 with fluoropolymers used in electronic devices and in microchips. There has also been some speculation that CF_3SF_5 is associated with high voltage equipment created from SF_6 (a breakdown product of high voltage equipment) reacting with CF_3

to form the CF_3SF_5 molecule. CF_3 is found as a breakdown product in fluoropolymers in this type of equipment. Furthermore, the rate of concentration increases of CF_3SF_5 tracks very closely to the trend observed in atmospheric SF_6 levels, indicating that these two molecules are connected.

Pepi et al. 2005 investigated the possibility of CF_3SF_5 through reactions of SF_6 in high voltage equipment and electrical devices. In their experiments, they attempted to react SF_6 with various fluoropolymers containing CF_4, C_2F_6, and C3F$_8$, stating that these reactions lend the possibility of forming CF_3SF_5. However, after mass spectrometric analysis of the resulting gas mixture, no detectable levels of CF_3SF_5 were found. The conclusion of the authors is that CF_3SF_5 is not produced from reactions of SF_6 and fluoropolymers used in high voltage equipment.

The source of CF_3SF_5 remains somewhat of a mystery. The hypothesis that CF_3SF_5 may be a by-product of electrical equipment processes seems to have failed based on the investigations conducted by Pepi et al. The only seemingly definitive source of CF_3SF_5 was through the formation of fluoropolymers presented by Santoro 2000. No levels of production or chemical releases were provided, so it is unknown whether this process is the primary contributor to current atmospheric levels.

References

- "Rising Ozone Levels Pose Challenge to U.S. Soybean Production, Scientists Say". NASA Earth Observatory. 2003-07-31. Retrieved 2006-05-10

- Emit-greenhouse-gas-nitrous-oxide, news: phys.org, Retrieved 10 February, 2020

- Dahl, R (2006). "Ozone Overload: Current Standards May Not Protect Health". Environ. Health Perspect. 114 (4): A240. Doi:10.1289/ehp.114-a240a. PMC 1440818

- Brown, Theodore L.; lemay, H. Eugene, Jr.; Bursten, Bruce E.; Burdge, Julia R. (2003) [1977]. "22". In Nicole Folchetti (ed.). Chemistry: The Central Science (9th ed.). Pearson Education. Pp. 882–883. ISBN 978-0-13-066997-1

- Turner, M. C., Jerrett, M., Pope III, C. A., Krewski, D., Gapstur, S. M., Diver, W. R., ... & Burnett, R. T. (2016). Long-term ozone exposure and mortality in a large prospective study. American journal of respiratory and critical care medicine, 193(10), 1134-1142

- "Methane in the atmosphere is surging, and that's got scientists worried". Latimes.com. 1 March 2019. Retrieved 1 March 2020

- Nazaries, Loïc; et al. (September 2013). "Methane, microbes and models: fundamental understanding of the soil methane cycle for future predictions". Environmental Microbiology. 15 (9): 2395–2417. Doi:10.1111/1462-2920.12149. PMID 23718889

- Gale, Joseph (2009). Astrobiology of Earth : the emergence, evolution, and future of life on a planet in turmoil. Oxford: Oxford University Press. ISBN 978-0-19-920580-6

- Loïc Jounot (2006). "Tropospheric Chemistry". University of Toronto Atmospheric Physics Department. Archived from the original on 17 June 2008. Retrieved 2008-07-18

Negative Impacts of Greenhouse Gas

Greenhouse gases are responsible for affecting the environment adversely in many ways. Greenhouse effect, runaway greenhouse effect, enhanced greenhouse effect, global warming, radiative forcing, etc. are some of these effects. This chapter has been carefully written to provide an easy understanding of the various effects of greenhouse gases.

GREENHOUSE EFFECT

The greenhouse effect is the rise in temperature that the Earth experiences because certain gases in the atmosphere (water vapor, carbon dioxide, nitrous oxide, ozone, methane, for example) trap energy that comes from the sun. These gases are usually called greenhouse gases since they behave much like the glass panes in a greenhouse. The glass panels of the greenhouse let in the light but keep heat from escaping and this is similar to the effect these gasses have on earth.

Sunlight enters the Earth's atmosphere, passing through the greenhouse gases. As it reaches the Earth's surface, land, water, and biosphere absorb the sunlight's energy. Once absorbed, this energy is sent back into the atmosphere. Some of the energy passes back into space, but much of it remains trapped in the atmosphere by the greenhouse gases. This is the completely natural process and without these gases all the heat would escape back into space and Earth's average temperature would be about 30 degrees Celsius (54 degrees Fahrenheit) colder. The greenhouse effect is very important process, because without the greenhouse effect, the Earth would not be warm enough for humans to live. But if the greenhouse effect becomes stronger, it could make the Earth warmer than usual. Even a little extra warming may cause problems for humans, plants, and animals.

The Enhanced Greenhouse Effect

Some human activities also produce greenhouse gases and these gases keep increasing in the atmosphere. The change in the balance of the greenhouse gases has significant effects on the entire planet. Burning fossil fuels - coal, oil and natural gas - releases carbon dioxide into the atmosphere. Cutting down and burning trees also produces a lot of carbon dioxide. A group of greenhouse gases called the chlorofluorocarbons have been used in aerosols, such as hairspray cans, fridges and in making foam plastics.

Since there are more and more greenhouse gases in the atmosphere, more heat is trapped, which makes the Earth warmer. This is known as global warming. A lot of scientists agree that man's activities are making the natural greenhouse effect stronger. If we carry on polluting the atmosphere with greenhouse gases, it will have very dangerous effects on the Earth. Today, the increase in the Earth's temperature is increasing with unprecedented speed.

To understand just how quickly global warming is accelerating, consider that during the entire 20th century, the average global temperature increased by about 0.6 degrees Celsius (slightly more than 1 degree Fahrenheit). Using computer climate models, scientists estimate that by the year 2100 the average global temperature will increase by 1.4 degrees to 5.8 degrees Celsius (approximately 2.5 degrees to 10.5 degrees Fahrenheit).

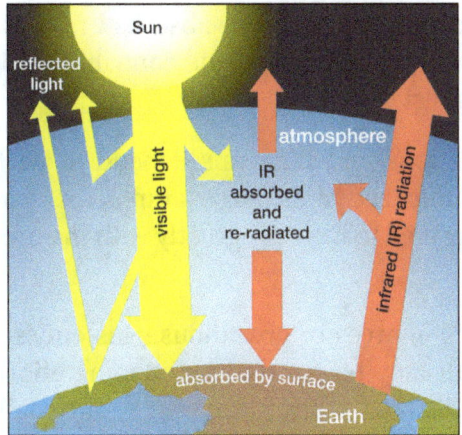

The greenhouse effect.

Greenhouse Gases

Many greenhouse gases occur naturally in the environment, such as water vapor, carbon dioxide, methane, nitrous oxide, and ozone. Others such as hydrofluorocarbons (HFCs), perfluorocarbons (PFCs), and sulfur hexafluoride (SF_6) are created and emitted solely through human activities. Human activities also add significantly to the level of naturally occurring greenhouse gases. The principal greenhouse gases that enter the atmosphere because of human activities are:

- Carbon Dioxide (CO_2): Carbon dioxide enters the atmosphere through the burning of fossil fuels (oil, natural gas, and coal), solid waste, trees and wood products, and also as a result of other chemical reactions (e.g., manufacture of cement). Carbon dioxide is also removed from the atmosphere (or "sequestered") when it is absorbed by plants as part of the biological carbon cycle.

- Nitrous Oxide (N_2O): Nitrous oxide is emitted during various agricultural and industrial activities, as well as during combustion of fossil fuels and solid waste.

- Methane (CH_4): Methane is emitted during the production and transport of coal, natural gas, and oil. Methane is also emitted when organic waste decomposes, whether in landfills or in connection with livestock farming.

- Fluorinated Gases: Hydrofluorocarbons, perfluorocarbons, and sulfur hexafluoride are synthetic, powerful greenhouse gases that are emitted from a variety of industrial processes. Fluorinated gases are sometimes used as substitutes for ozone-depleting substances (i.e., CFCs, HCFCs, and halons). These gases are typically emitted in smaller quantities, but because they are potent greenhouse gases, they are sometimes referred to as High Global Warming Potential gases ("High GWP gases").

Greenhouse gases vary in their ability to absorb and hold heat in the atmosphere. HFCs and PFCs are the most heat-absorbent, but there are also wide differences between naturally occurring gases. For example, nitrous oxide absorbs 270 times more heat per molecule than carbon dioxide, and methane absorbs 21 times more heat per molecule than carbon dioxide. However, carbon dioxide contributes the most, since its level in the atmosphere is the highest.

Estimates of future emissions and removals depend in part on assumptions about changes in underlying human activities. For example, the demand for fossil fuels such as gasoline and coal is expected to increase greatly with the predicted growth of the U.S. and global economies.

Many, but not all, human sources of greenhouse gas emissions are expected to rise in the future. This growth may be reduced by ongoing efforts to increase the use of newer, cleaner technologies and other measures. Additionally, our everyday choices about such things as commuting, housing, electricity use, and recycling can influence the amount of greenhouse gases being emitted.

The effects of Global Warming

With more heat trapped on Earth, the planet will become warmer, which means the weather all over Earth will change. Since the conditions we are living in are perfect for life, a large rise in temperature could be disastrous for us and for any other living creatures on Earth. At the moment, it is difficult for scientists to say how big the changes will be and where the worst effects will occur. These are some of the assumptions.

The Weather

The effects will vary in different parts of the world: some places will become drier and others will become wetter. Although most areas will be warmer, some areas will become cooler. There may be many storms, floods and drought, but we do not know which areas of the world will be affected. All over the world, these weather changes will affect the kinds of crop that can be grown. Plants, animals, and even people may find it difficult to survive in different conditions.

Sea Levels

Higher temperatures will make the water of the seas and oceans expand. Ice melting in the Antarctic and Greenland will flow into the sea. All over the world, sea levels may rise, perhaps by as much as 20 to 40 cm, by the beginning of the next century. Higher sea levels will threaten the low-lying coastal areas of the world, such as the Netherlands and Bangladesh. Throughout the world, millions of people and areas of land will be at danger from flooding. Many people will have to leave their homes and large areas of farmland will be ruined because of floods.

Farming

The changes in the weather will affect the types of crops grown in different parts of the world. Some crops, such as wheat and rice, grow better in higher temperatures, but other plants, such as maize and sugarcane, do not. Changes in the amount of rainfall will also affect how many plants grow. The effect of a change in the weather on plant growth may lead to some countries not having enough food. Brazil, parts of Africa, south-east Asia, and China will be affected the most and many people could suffer from hunger.

Plants and Animals

It has taken millions of years for life to become used to the conditions on Earth. As weather and temperature changes, the homes of plants and animals will be affected all over the world. For example, polar bears and seals will have to find new land for hunting and living if the ice in the Arctic melts. Many animals and plants may not be able to cope with these changes and could die. This could cause the loss of some animal and plant species in certain or all areas of the world.

People

The changes in climate will affect everyone, but some populations will be at greater risk. For example, countries whose coastal regions have a large population, such as Egypt and China, may see whole populations move inland to avoid flood risk areas. The effect on people will depend on how well we can adapt to the changes and how much we can do to reduce climate change in the world.

Relationship between Climate Change and Global Public Health

Consensus exists among scientists all over the globe that the world's climate is changing and that these changes can affect human health. The more direct health effects of climate change can include injuries and illnesses from severe weather, floods, and heat exposure; increases in disease caused by allergies, respiratory problems, and illnesses carried by insects or in water; and threats to the safety and availability of our food and water supplies. Less direct effects can include worry, depression, and the negative impacts of mass migration and regional conflicts.

To a large extent, public health depends on safe drinking water, sufficient food, secure shelter, and good social conditions. A changing climate is likely to affect all of these conditions. Warming climate as a result of the greenhouse effect is likely to bring some localized benefits, such as decreased winter deaths in temperate climates, and increases in food production in some regions.

However, the health effects of a rapidly changing climate are likely to be overwhelmingly negative, particularly in the poorest communities, which have contributed least to greenhouse gas emissions. Some of the health effects include increase in frequencies of heatwaves, shortages in supplies of freshwater, rise in temperatures followed by variable precipitation, which are likely to decrease the production of staple foods in many of the poorest regions, rising sea levels, and prolongation of seasons for transmission of important vector-borne disease, as well as the alteration of their geographical range. All these events may lead to increased risks of:

- Water-borne disease.

- Malnutrition.

- Coastal flooding.

- Huge population displacement.

- New diseases moving into the regions which lack either population immunity or a strong public health infrastructure.

Measurement of health effects from climate change can only be very approximate. Nevertheless, a WHO quantitative assessment, taking into account only a subset of the possible health impacts, concluded that the effects of the climate change that has occurred since the mid-1970s may have caused over 150,000 deaths in 2000. It also concluded that these impacts are likely to increase in the future.

So far not many measures have been taken to address climate change. This is largely caused by the major uncertainties still surrounding the theory. But climate change is also a global problem that is hard to solve by single countries. Therefore in 1998, the Kyoto Protocol was negotiated in Kyoto, Japan. It requires participating countries to reduce their anthropogenic greenhouse gas emissions (CO_2, CH_4, N_2O, HFCs, PFCs, and SF_6) by at least 5% below 1990 levels in the commitment period 2008 to 2012. The Kyoto Protocol was eventually signed in Bonn in 2001 by 186 countries. Several countries such as the United States and Australia have retreated.

Runaway Greenhouse Effect

A runaway greenhouse effect is when there is enough of a greenhouse gas in a planet's atmosphere such that the gas blocks thermal radiation from the planet, preventing the planet from cooling and from having liquid water on its surface. The runaway greenhouse effect can be defined by a limit on a planet's outgoing longwave radiation which is asymptotically reached due to higher surface temperatures boiling a condensable species (often water vapor) into the atmosphere, increasing its optical depth. This runaway positive feedback means the planet cannot cool down through longwave radiation (via the Stefan–Boltzmann law) and continues to heat up until it can radiate outside of the absorption bands of the condensable species.

The runaway greenhouse effect is often formulated with water vapor as the condensable species. In this case the water vapor reaches the stratosphere and escapes into space via hydrodynamic escape, resulting in a desiccated planet.

Physics of the Runaway Greenhouse

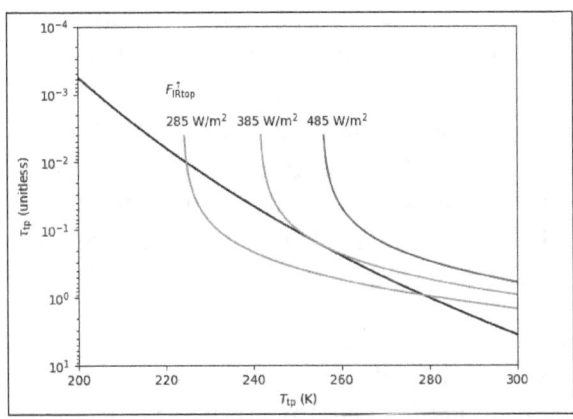

Graph of tropopause optical depth by tropopause temperature, illustrating the Komabayashi-Ingersoll limit of 385 W/m^2 using equations and values from Nakajima et al. "A Study on the Runaway Greenhouse Effect with a One-Dimensional Radiative–Convective Equilibrium Model". The Komabayashi-Ingersoll limit is the value of outgoing longwave radiation (FIRtop) beyond which the lines do not intersect.

The runaway greenhouse effect is often formulated in terms of how the surface temperature of a planet changes with differing amounts of received starlight. If the planet is assumed to be in radiative equilibrium, then the runaway greenhouse state is calculated as the equilibrium state at which water cannot exist in liquid form. The water vapor is then lost to space through hydrodynamic escape. In radiative equilbrium, a planet's outgoing longwave radiation (OLR) must balance the incoming stellar flux. Typically a planet's outgoing longwave radiation is equal to the planet's surface temperature to the fourth power due to the Stefan–Boltzmann law, though in a runaway greenhouse state this balance is broken.

The Stefan-Boltzmann law is an example of a negative feedback that stabilizes a planet's climate system. If the Earth received more sunlight it would result in a temporary disequilibrium (more energy in than out) and result in warming. However, because the Stefan-Boltzmann response mandates that this hotter planet emits more energy, eventually a new radiation balance can be reached and the temperature will be maintained at its new, higher value. Positive climate change feedbacks amplify changes in the climate system, and can lead to destabilizing effects for the climate. An increase in temperature from greenhouse gases leading to increased water vapor (which is itself a greenhouse gas) causing further warming is a positive feedback, but not a runaway effect, on Earth. Positive feedback effects are common (e.g. ice-albedo feedback) but runaway effects do not necessarily emerge from their presence. Though water plays a major role in the process, the runaway greenhouse effect is not a result of water vapor feedback.

The runaway greenhouse effect can be seen as a limit on a planet's outgoing longwave radiation that, when surpassed, results in a state where water cannot exist in its liquid form (hence, the oceans have all "boiled away"). A planet's outgoing longwave radiation is limited by this evaporated water, which is an effective greenhouse gas and blocks additional infrared radiation as it accumulates in the atmosphere. Assuming radiative equilibrium, runaway greenhouse limits on outgoing longwave radiation correspond to limits on the increase in stellar flux received by a planet to trigger the runaway greenhouse effect. Two limits on a planet's outgoing longwave radiation have been calculated that correspond with the onset of the runaway greenhouse effect: the Komabayashi-Ingersoll limit and the Simpson-Nakajima limit. At these values the runaway greenhouse effect overcomes the Stefan-Boltzmann feedback so an increase in a planet's surface temperature will not increase the outgoing longwave radiation.

The Komabayashi-Ingersoll limit was the first to be analytically derived and only considers a grey stratosphere in radiative equilibrium. A grey stratosphere (or atmosphere) is an approach to modeling radiative transfer that does not take into account the frequency-dependence of absorption by a gas. In the case of a grey stratosphere or atmosphere, the Eddington approximation can be used to calculate radiative fluxes. This approach focuses on the balance between the outgoing longwave radiation at the tropopause, $F_{IRtop}^{\uparrow}$, and the optical depth of water vapor, τ_{tp}, in the tropopause, which is determined by the temperature and pressure at the tropopause according to the saturation vapor pressure. This balance is represented by the following equations:

$$\frac{1}{2} F_{IRtop}^{\uparrow} \left(\frac{3}{2} \tau_{tp} + 1 \right) = \sigma T_{tp}^{4}$$

$$\tau_{tp} = \kappa_{v} p^{*} (T_{tp}) \frac{1}{g} \frac{m_{v}}{\overline{m}}$$

Where the first equation represents the requirement for radiative equilibrium at the tropopause and the second equation represents how much water vapor is present at the tropopause. Taking the outgoing longwave radiation as a free parameter, these equations will intersect only once for a single value of the outgoing longwave radiation, this value is taken as the Komabayashi-Ingersoll limit. At that value the Stefan-Boltzmann feedback breaks down because the tropospheric temperature required to maintain the Komabayashi-Ingersoll OLR value results in a water vapor optical depth that blocks the OLR needed to cool the tropopause.

The Simpson-Nakajima limit is lower than the Komabayashi-Ingersoll limit, and is thus typically more realistic for the value at which a planet enters a runaway greenhouse state. For example, given the parameters used to determine a Komabayashi-Ingersoll limit of $385 \, W/m^2$, the corresponding Simpson-Nakajima limit is only about $293 \, W/m^2$. The Simpson-Nakajima limit builds off of the derivation of the Komabayashi-Ingersoll limit by assuming a convective troposphere with a surface temperature and surface pressure that determines the optical depth and outgoing longwave radiation at the tropopause.

The Moist Greenhouse Limit

Because the model used to derive the Simpson-Nakajima limit (a grey stratosphere in radiative equilibrium and a convecting troposphere) can determine the water concentration as a function of altitude, the model can also be used to determine the surface temperature (or conversely, amount of stellar flux) that results in a high water mixing ratio in the stratosphere. While this critical value of outgoing longwave radiation is less than the Simpson-Nakajima limit, it still has dramatic effects on a planet's climate. A high water mixing ratio in the stratosphere would overcome the effects of a cold trap and result in a "moist" stratosphere, which would result in the photolysis of water in the stratosphere that in turn would destroy the ozone layer and eventually lead to a dramatic loss of water through hydrodynamic escape. This climate state has been dubbed the moist greenhouse effect, as the end-state is a planet without water, though liquid water may exist on the planet's surface during this process.

Connection to Habitability

The concept of a habitable zone has been used by planetary scientists and astrobiologists to define an orbital region around a star in which a planet (or moon) can sustain liquid water. Under this definition, the inner edge of the habitable zone (i.e., the closest point to a star that a planet can be until it can no longer sustain liquid water) is determined by the outgoing longwave radiation limit beyond which the runaway greenhouse process occurs (e.g., the Simpson-Nakajima limit). This is because a planet's distance from its host star determines the amount of stellar flux the planet receives, which in turn determines the amount of outgoing longwave radiation the planet radiates back to space. While the inner habitable zone is typically determined by using the Simpson-Nakajima limit, it can also be determined with respect to the moist greenhouse limit, though the difference between the two is often small.

Calculating the inner edge of the habitable zone is strongly dependent on the model used to calculate the Simpson-Nakajima or moist greenhouse limit. The climate models used to calculate these limits have evolved over time, with some models assuming a simple one-dimensional, grey atmosphere, and others using a full radiative transfer solution to model the absorption bands of water and carbon dioxide. These earlier models that used radiative transfer derived the absorption coefficients for water from the HITRAN database, while newer models use the more current and accurate HITEMP database, which has led to different calculated values of thermal radiation limits. More accurate calculations have been done using three-dimensional climate models that take into account effects such as planetary rotation and local water mixing ratios as well as cloud feedbacks. The effect of clouds on calculating thermal radiation limits is still in debate (specifically, whether or not water clouds present a positive or negative feedback effect).

Earth

Early investigations on the effect of atmospheric carbon dioxide levels on the runaway greenhouse limit found that it would take orders of magnitude higher amounts of carbon dioxide to take the Earth to a runaway greenhouse state. This is because carbon dioxide is not anywhere near as effective at blocking outgoing longwave radiation as water is. Within current models of the runaway greenhouse effect, carbon dioxide (especially anthropogenic carbon dioxide) does not seem capable of providing the necessary insulation for the Earth to reach the Simpson-Nakajima limit.

There remains debate, however, on whether carbon dioxide can push surface temperatures towards the moist greenhouse limit. Climate scientist John Houghton has written that "(there) is no possibility of (Venus's) runaway greenhouse conditions occurring on the Earth". The IPCC (Intergovernmental Panel on Climate Change) has also stated that "a 'runaway greenhouse effect'—analogous to (that of) Venus—appears to have virtually no chance of being induced by anthropogenic activities." However, climatologist James Hansen disagrees. In his Storms of My Grandchildren he says that burning coal and mining oil sands will result in runaway greenhouse on Earth. A re-evaluation in 2013 of the effect of water vapor in the climate models showed that James Hansen's outcome might be possible, but requires ten times the amount of CO_2 we could release from burning all the oil, coal, and natural gas in Earth's crust. As with the uncertainties in calculating the inner edge of the habitable zone, the uncertainty in whether CO_2 can drive a moist greenhouse effect is due to differences in modeling choices and the uncertainties therein. The switch from using HITRAN to the more current HITEMP absorption line lists in radiative transfer calculations has shown that previous runaway greenhouse limits were too high, but the necessary amount of carbon dioxide would make an anthropogenic moist greenhouse state unlikely. Full three-dimensional models have shown that the moist greenhouse limit on surface temperature is higher than that found in one-dimensional models and thus would require a higher amount of carbon dioxide to initiate a moist greenhouse than in one-dimensional models. Other complications include whether the atmosphere is saturated or sub-saturated at some humidity, higher CO_2 levels in the atmosphere resulting in a less hot Earth than expected due to Rayleigh scattering, and whether cloud feedbacks stabilize or destabilize the climate system.

Complicating the matter, research on Earth's climate history has often used the term run-away greenhouse effect to describe large-scale climate changes when it is not an appropriate description as it does not depend on Earth's outgoing longwave radiation. Though the Earth has experienced a diversity of climate extremes, these are not end-states of climate evolution and have instead represented climate equilibria different than that seen on Earth today. For example, it has been hypothesized that large releases of greenhouse gases may have occurred concurrently with the Permian–Triassic extinction event or Paleocene–Eocene Thermal Maximum. Additionally, during 80% of the latest 500 million years, the Earth is believed to have been in a greenhouse state due to the greenhouse effect, when there were no continental glaciers on the planet, the levels of carbon dioxide and other greenhouse gases (such as water vapor and methane) were high, and sea surface temperatures (SSTs) ranged from 28 °C (82.4 °F) in the tropics to 0 °C (32 °F) in the polar regions. Other terms, such as "abrupt climate change", or tipping points could be used when describing such scenarios, as well as the terms hothouse state or greenhouse earth.

Distant Future

Most scientists believe that a runaway greenhouse effect is actually inevitable in the long term as the Sun gradually gets bigger and hotter as it ages. Such will potentially spell the end of all life on Earth. As the Sun becomes 10% brighter in about one billion years' time, the surface temperature of Earth will reach 47 °C (117 °F), causing the temperature of Earth to rise rapidly and its oceans to boil away until it becomes a greenhouse planet similar to Venus today.

According to astrobiologists Peter Ward and Donald Brownlee in their book The Life and Death of Planet Earth, the current loss rate is approximately one millimeter of ocean per million years, but this rate is gradually accelerating as the sun gets warmer, to perhaps as fast as one millimeter every 1000 years. Ward and Brownlee predict that there will be two variations of this future warming feedback: the "moist greenhouse" where water vapor dominates the troposphere and starts to accumulate in the stratosphere, and the "runaway greenhouse" where water vapor becomes a dominant component of the atmosphere such that the Earth starts to undergo rapid warming that could send its surface temperature to over 900 °C (1,650 °F), causing its entire surface to melt and killing all life, perhaps in about three billion years' time. In both the moist and runaway greenhouse states the loss of oceans will turn the Earth into a primarily desert world. The only water left on the planet would be in a few evaporating ponds scattered near the poles as well as huge salt flats around what was once the ocean floor, much like the Atacama Desert in Chile or Badwater Basin in Death Valley. These small reservoirs of water may allow life to remain for a few billion more years.

As the Sun brightens, CO_2 levels should decrease due to an increase of activity in the carbon-silicate cycle corresponding to the increase of temperature. This would mitigate some of the heating Earth would experience due to the Sun's increase in brightness. Eventually though, as the water escapes, the carbon cycle will cease as plate tectonics come to a halt due to the need for water as a lubricant for tectonic activity.

GLOBAL WARMING

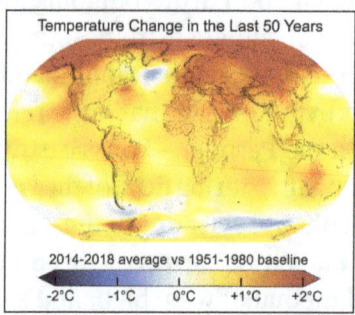

Average global temperatures from 2014 to 2018 compared to a baseline average from 1951 to 1980, according to NASA's Goddard Institute for Space Studies.

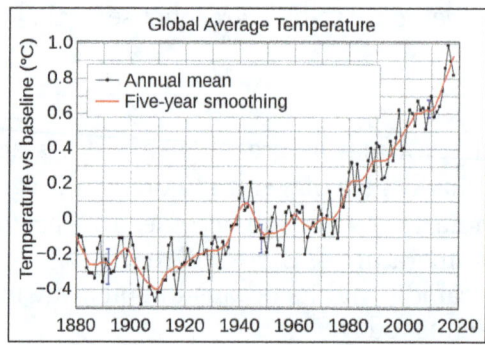

The average annual temperature at the earth's surface has risen since the late 1800s, with year-to-year variations (shown in black) being smoothed out (shown in red) to show the general warming trend.

Global warming is the long-term rise in the average temperature of the Earth's climate system. It is a major aspect of current climate change, and has been demonstrated by direct temperature measurements and by measurements of various effects of the warming. The term commonly refers to the mainly human-caused increase in global surface temperatures and its projected continuation. In this context, the terms global warming and climate change are often used interchangeably, but climate change includes both global warming and its effects, such as changes in precipitation and impacts that differ by region. There were prehistoric periods of global warming, but observed changes since the mid-20th century have been much greater than those seen in previous records covering decades to thousands of years.

The Intergovernmental Panel on Climate Change (IPCC) Fifth Assessment Report concluded, "It is extremely likely that human influence has been the dominant cause of the observed warming since the mid-20th century." The largest human influence has been the emission of greenhouse gases such as carbon dioxide, methane, and nitrous oxide. Climate model projections summarized in the report indicated that during the 21st century the global surface temperature is likely to rise a further 0.3 to 1.7 °C (0.5 to 3.1 °F) in a moderate scenario, or as much as 2.6 to 4.8 °C (4.7 to 8.6 °F) in an extreme scenario, depending on the rate of future greenhouse gas emissions and on climate feedback effects. These findings have been recognized by the national science academies of the major industrialized nations and are not disputed by any scientific body of national or international standing.

The effects of global warming include rising sea levels, regional changes in precipitation, more frequent extreme weather events such as heat waves, and expansion of deserts. Surface temperature increases are greatest in the Arctic, which has contributed to the retreat of glaciers, permafrost, and sea ice. Overall, higher temperatures bring more rain and snowfall, but for some regions droughts and wildfires increase instead. Climate change threatens to diminish crop yields, harming food security, and rising sea levels may flood coastal infrastructure and force the abandonment of many coastal cities. Environmental impacts include the extinction or relocation of many species as their ecosystems change, most immediately the environments of coral reefs, mountains, and the Arctic. Due to the persistence of CO_2 in the atmosphere and the inertia of the climate system, climatic changes and their effects will continue for millennia even if carbon emissions are stopped.

Possible societal responses to global warming include mitigation by emissions reduction, adaptation to its effects, and maybe climate engineering. Countries work together on climate change under the umbrella of the United Nations Framework Convention on Climate Change (UNFCCC), which has near-universal membership. The ultimate goal of the convention is to "prevent dangerous anthropogenic interference with the climate system". Although the parties to the UNFCCC have agreed that deep cuts in emissions are required and that global warming should be limited to well below 2 °C (3.6 °F) in the Paris Agreement, the Earth's average surface temperature has already increased by about half this threshold and current pledges by countries to cut emissions are inadequate to limit future warming.

Observed Temperature Changes

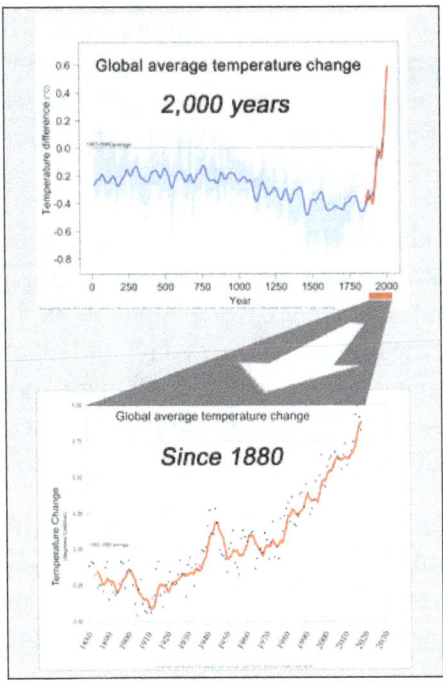

Global average temperatures declined for thousands of years, until fossil fuel-based industrialization beginning roughly 200 years ago reversed the decline. Global warming has intensified in recent decades.

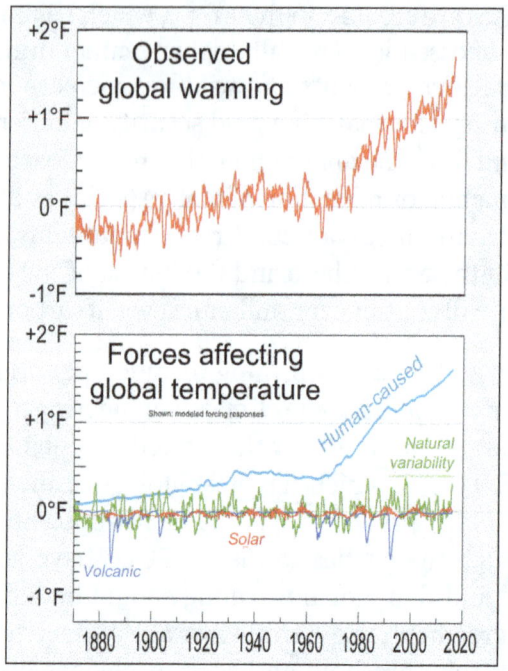

Scientists have investigated many possible causes of global warming, and have found that accumulation in the atmosphere of greenhouse gases, especially those resulting from humans burning fossil fuels, is the predominant cause.

Climate proxy records show that natural variations offset the early effects of the Industrial Revolution, so there was little net warming between the 18th century and the mid-19th century, when thermometer records began to provide global coverage. The IPCC has adopted the baseline reference period 1850–1900 as an approximation of pre-industrial global mean surface temperature.

Multiple independently produced instrumental datasets confirm that the 2009–2018 decade was 0.93 ± 0.07 °C warmer than the pre-industrial baseline. Currently, surface temperatures are rising by about 0.2 °C per decade. Since 1950, the number of cold days and nights have decreased, and the number of warm days and night have increased. Historical patterns of warming and cooling, like the Medieval Climate Anomaly and the Little Ice Age, were not as synchronous as current warming, but may have reached temperatures as high as those of the late-20th century in a limited set of regions.

Although the most common measure of global warming is the increase in the near-surface atmospheric temperature, over 90% of the additional energy stored in the climate system over the last 50 years has warmed ocean water. The remainder of the additional energy has melted ice and warmed the continents and the atmosphere.

The warming evident in the instrumental temperature record is consistent with a wide range of observations, documented by many independent scientific groups; for example, in most continental regions the frequency and intensity of heavy precipitation has increased. Further examples include sea level rise, widespread melting of snow and land ice, increased heat content of the oceans, increased humidity, and the earlier timing of spring events, such as the flowering of plants.

Regional Trends

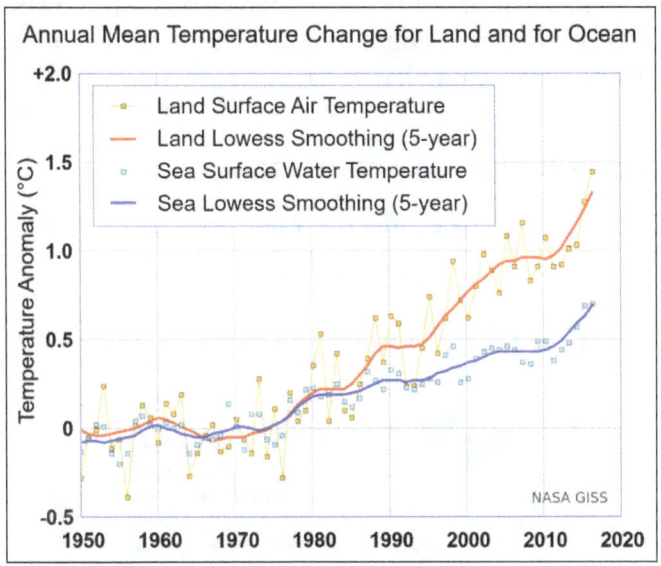

Average annual temperature has risen faster on
land than on the ice-free surface of the sea.

Global warming refers to global averages, with the amount of warming varying by region. Since the pre-industrial period, global average land temperatures have increased almost twice as fast as global average temperatures. This is due to the larger heat capacity of oceans and because oceans lose more heat by evaporation. Patterns of warming are independent of the locations of greenhouse gas emissions because the gases persist long enough to diffuse across the planet; however, localized black carbon deposits on snow and ice do contribute to Arctic warming.

The Northern Hemisphere and North Pole have warmed much faster than the South Pole and Southern Hemisphere. The Northern Hemisphere not only has much more land, but the arrangement of land masses around the Arctic Ocean has resulted in the maximum surface area flipping from reflective snow and ice cover to ocean and land surfaces that absorb more sunlight and thus more heat. Arctic temperatures have increased and are predicted to continue to increase during this century at over twice the rate of the rest of the world. As the temperature difference between the Arctic and the equator decreases, ocean currents that are driven by that temperature difference, like the Gulf Stream, are weakening.

Short-term Slowdowns and Surges

Because the climate system has large thermal inertia, it can take centuries for the climate to fully adjust. While record-breaking years attract considerable public interest, individual years are less significant than the overall trend. Global surface temperature is subject to short-term fluctuations that overlie long-term trends, and can temporarily mask or magnify them. An example of such an episode is the slower rate of surface temperature increase from 1998 to 2012, which was dubbed the global warming hiatus. Throughout this period ocean heat storage continued to progress steadily upwards, and in subsequent years surface temperatures have spiked upwards. The slower pace of warming can be attributed to a combination of natural fluctuations, reduced solar activity, and increased volcanic activity.

Physical Drivers of Recent Climate Change

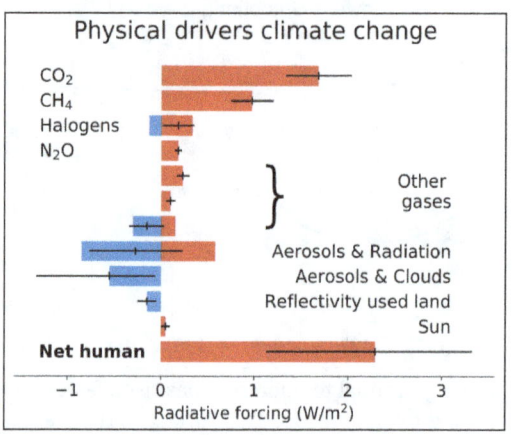

Radiative forcing of different contributors to climate change in 2011, as reported in the fifth IPCC assessment report. For the gases and aerosols, the values represent both the effect they have themselves and the effect of any chemical compound they get converted into in the atmosphere.

By itself, the climate system experiences various cycles which can last for years (such as the El Niño–Southern Oscillation) to decades or centuries. Other changes are caused by external forcings. These forcings are "external" to the climate system, but not always external to the Earth. Examples of external forcings include changes in the composition of the atmosphere (e.g. increased concentrations of greenhouse gases), solar luminosity, volcanic eruptions, and variations in the Earth's orbit around the Sun.

Attributing detected temperature changes and extreme events to human-caused increases in greenhouse gases requires scientists to rule out known internal climate variability and natural external forcings. Therefore, a key approach is to use physically or statistically based computer modelling of the climate system to determine unique fingerprints for all potential causes. By comparing these fingerprints with observed patterns and evolution of climate change, and the observed evolution of the forcings, the causes of the observed changes can be determined. Scientists have determined that the major factors causing the current climate change are greenhouse gases, land use changes, and aerosols and soot.

Greenhouse Gases

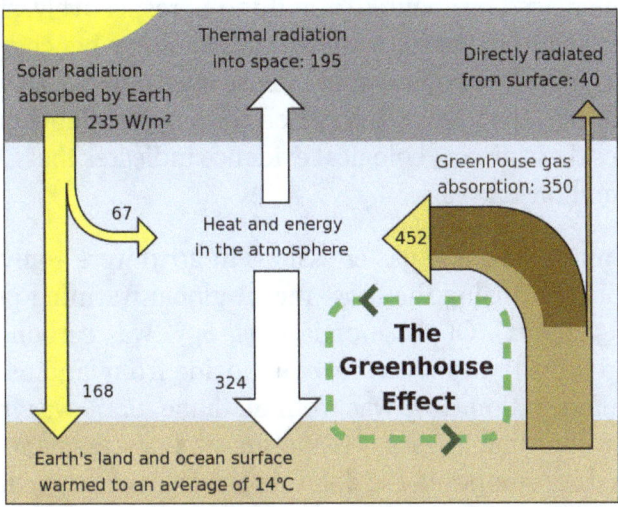

Greenhouse effect schematic showing energy flows between space, the atmosphere, and the Earth's surface. Energy exchanges are expressed in watts per square meter (W/m²).

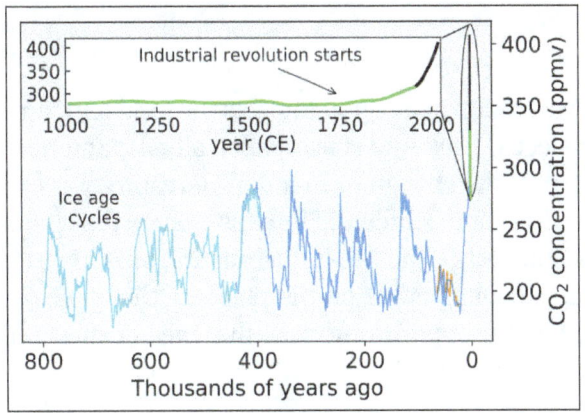

CO_2 concentrations over the last 800,000 years as measured from ice cores (blue/green) and directly (black).

Greenhouse gases trap heat radiating from the Earth to space. This heat, in the form of infrared radiation, gets absorbed and emitted by these gases in the atmosphere, thus warming the lower atmosphere and the surface. Before the Industrial Revolution, naturally occurring amounts of greenhouse gases caused the air near the surface to be warmer by about 33 °C (59 °F) than it would be in their absence. Without the Earth's atmosphere, the Earth's average temperature would be well below the freezing temperature of water. While water vapour (~50%) and clouds (~25%) are the biggest contributors to the greenhouse effect, they increase as a function of temperature and are therefore considered feedbacks. Increased concentrations of gases such as CO_2 (~20%), ozone and N_2O are external forcing on the other hand.

Human activity since the Industrial Revolution has increased the amount of greenhouse gases in the atmosphere, leading to increased radiative forcing from CO_2, methane, tropospheric ozone, CFCs, and nitrous oxide. As of 2011, the concentrations of CO_2 and methane had increased by about 40% and 150%, respectively, since pre-industrial times. In 2013, CO_2 readings taken at the world's primary benchmark site in Mauna Loa surpassing 400 ppm for the first time. These levels are much higher than at any time during the last 800,000 years, the period for which reliable data have been collected from ice cores. Less direct geological evidence indicates that CO_2 values have not been this high for millions of years.

Global anthropogenic greenhouse gas emissions in 2010 were equivalent to 49 billion tonnes of carbon dioxide (using the most recent global warming potentials over 100 years from the AR5 report). Of these emissions, 65% was carbon dioxide from fossil fuel burning and industry, 11% was carbon dioxide from land use change, which is primarily due to deforestation, 16% was from methane, 6.2% was from nitrous oxide, and 2.0% was from fluorinated gases. Using life-cycle assessment to estimate emissions relating to final consumption, the dominant sources of 2010 emissions were: food (26–30% of emissions); washing, heating, and lighting (26%); personal transport and freight (20%); and building construction (15%).

Land use Change

Changing the type of vegetation in a region impacts the local temperature by changing how much sunlight gets reflected back into space, called albedo, and how much heat is lost by evaporation. For instance, the change from a dark forest to grassland makes the surface lighter, causing it to reflect more sunlight. Humans change the land surface mainly to create more agricultural land. Since the pre-industrial era, albedo has increased due to land use change, which has a cooling effect on the planet. Other processes linked to land use change however have had the opposite effect, so that the net effect remains unclear.

Aerosols and Soot

Solid and liquid particles known as aerosols – from volcanoes, plankton, and human-made pollutants – reflect incoming sunlight, cooling the climate. From 1961 to 1990, a gradual reduction in the amount of sunlight reaching the Earth's surface was observed, a phenomenon popularly known as global dimming, typically attributed to aerosols from biofuel and fossil fuel burning. Aerosol removal by precipitation gives tropospheric aerosols an atmospheric lifetime of only about a week, while stratospheric aerosols can remain in the atmosphere for a few years. Globally, aerosols have been declining since 1990, removing some of the masking of global warming that they had been providing.

Ship tracks can be seen as lines in these clouds over the Atlantic Ocean
on the East Coast of the United States as an effect of aerosols.

In addition to their direct effect by scattering and absorbing solar radiation, aerosols have indirect effects on the Earth's radiation budget. Sulfate aerosols act as cloud condensation nuclei and thus lead to clouds that have more and smaller cloud droplets. These clouds reflect solar radiation more efficiently than clouds with fewer and larger droplets. This effect also causes droplets to be of more uniform size, which reduces the growth of raindrops and makes clouds more reflective to incoming sunlight. Indirect effects of aerosols are the largest uncertainty in radiative forcing.

While aerosols typically limit global warming by reflecting sunlight, black carbon in soot that falls on snow or ice can contribute to global warming. Not only does this increase the absorption of sunlight, it also increases melting and sea level rise. Limiting new black carbon deposits in the Arctic could reduce global warming by 0.2 °C by 2050. When soot is suspended in the atmosphere, it directly absorbs solar radiation, heating the atmosphere and cooling the surface. In areas with high soot production, such as rural India, as much as 50% of surface warming due to greenhouse gases may be masked by atmospheric brown clouds.

Minor forcings: The Sun and Short-lived Greenhouse Gases

As the Sun is the Earth's primary energy source, changes in incoming sunlight directly affect the climate system. Solar irradiance has been measured directly by satellites, and indirect measurements are available beginning in the early 1600s. There has been no upward trend in the amount of the Sun's energy reaching the Earth, so it cannot be responsible for the current warming. Physical climate models are also unable to reproduce the rapid warming observed in recent decades when taking into account only variations in solar output and volcanic activity. Another line of evidence for the warming not being due to the Sun is how temperature changes differ at different levels in the Earth's atmosphere. According to basic physical principles, the greenhouse effect produces warming of the lower atmosphere (the troposphere), but cooling of the upper atmosphere (the stratosphere). If solar variations were responsible for the observed warming, warming of both the troposphere and the stratosphere would be expected, but that has not been the case.

Ozone in the lowest layer of the atmosphere, the troposphere, is itself a greenhouse gas. Furthermore, it is highly reactive and interacts with other greenhouse gases and aerosols.

Climate Change Feedback

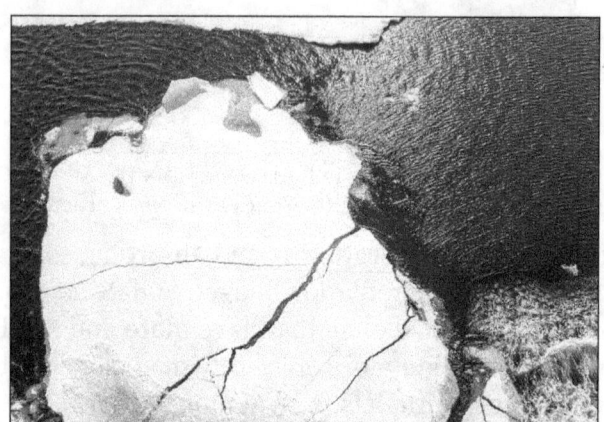

The dark ocean surface reflects only 6 percent of incoming solar radiation, whereas sea ice reflects 50 to 70 percent.

The response of the climate system to an initial forcing is increased by positive feedbacks and reduced by negative feedbacks. The main negative feedback to global temperature change is radiative cooling to space as infrared radiation, which increases strongly with increasing temperature. The main positive feedbacks are the water vapour feedback, the ice–albedo feedback, and probably the net effect of clouds. Uncertainty over feedbacks is the major reason why different climate models project different magnitudes of warming for a given amount of emissions.

As air gets warmer, it can hold more moisture. After an initial warming due to emissions of greenhouse gases, the atmosphere will hold more water. As water is a potent greenhouse gas, this further heats the climate: the water vapour feedback. The reduction of snow cover and sea ice in the Arctic reduces the albedo of the Earth's surface. More of the Sun's energy is now absorbed in these regions, contributing to Arctic amplification, which has caused Arctic temperatures to increase at more than twice the rate of the rest of the world. Arctic amplification also causes methane to be released as permafrost melts, which is expected to surpass land use changes as the second strongest anthropogenic source of greenhouse gases by the end of the century.

Cloud cover may change in the future. If cloud cover increases, more sunlight will be reflected back into space, cooling the planet. Simultaneously, the clouds enhance the greenhouse effect, warming the planet. The opposite is true if cloud cover decreases. It depends on the cloud type and location which process is more important. Overall, the net feedback over the industrial era has probably been positive. An analysis of satellite data between 1983 and 2009 reveals that cloud tops are reaching higher into the atmosphere and that cloudy storm tracks are shifting toward Earth's poles, suggesting clouds will be a positive feedback in the future.

Roughly half of each year's CO_2 emissions have been absorbed by plants on land and in oceans. Carbon dioxide and an extended growing season have stimulated plant growth making the land carbon cycle a negative feedback. Climate change also increases droughts and heat waves that inhibit plant growth, which makes it uncertain whether this negative feedback will persist in the future. Soils contain large quantities of carbon and may release some when they heat up. As more CO_2 and heat are absorbed by the ocean, it is acidifying and ocean circulation can change, changing the rate at which the ocean can absorb atmospheric carbon.

A concern is that positive feedbacks will lead to a tipping point, where global temperatures transition to a hothouse climate state even if greenhouse gas emissions are reduced or eliminated. A 2018 study tried to identify such a planetary threshold for self-reinforcing feedbacks and found that even a 2 °C (3.6 °F) increase in temperature over pre-industrial levels may be enough to trigger such a hothouse Earth scenario.

Climate Models

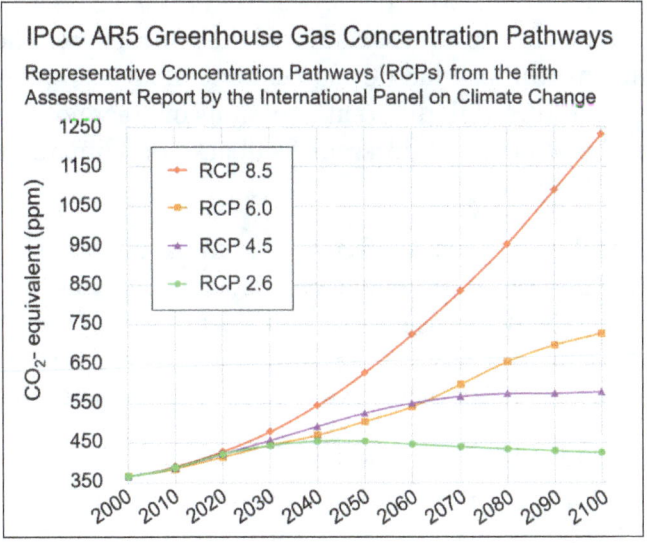

Future CO_2 projections, including all forcing agents' atmospheric CO_2-equivalent concentrations in parts-per-million-by-volume (ppmv) according to four RCPs (Representative Concentration Pathways).

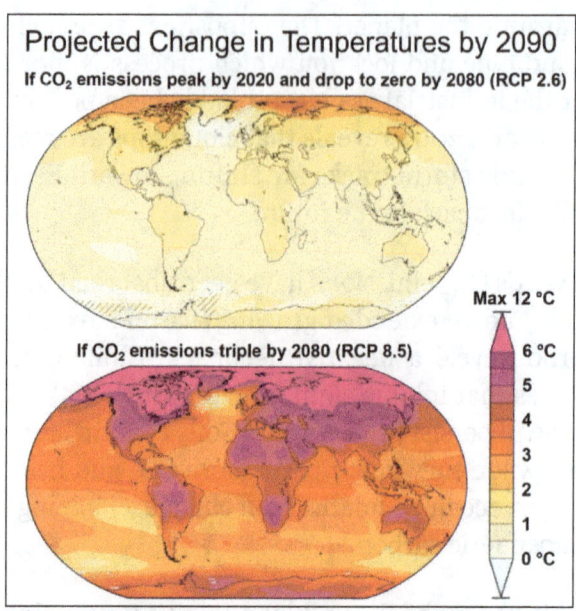

Projected Change in Temperatures by 2090

If CO_2 emissions peak by 2020 and drop to zero by 2080 (RCP 2.6)

If CO_2 emissions triple by 2080 (RCP 8.5)

Max 12 °C

6 °C
5
4
3
2
1
0 °C

Coupled Model Intercomparison Project Phase 5 (CMIP5) multi-model mean projections (i.e., the average of the model projections available) for the 2081–2100 period under the RCP 2.6 and RCP 8.5 scenarios for change in annual mean surface temperature. Changes are shown relative to the 1986–2005 period. Data from the IPCC Fifth Assessment Report.

A climate model is a representation of the physical, chemical, and biological processes that affect the climate system. Computer models are run on supercomputers to reproduce and predict the circulation of the oceans, the annual cycle of the seasons, and the flows of carbon between the land surface and the atmosphere. There are more than two dozen scientific institutions that develop climate models. Models not only project different future temperature with different emissions of greenhouse gases, but also do not fully agree on the strength of different feedbacks on climate sensitivity and the amount of inertia of the system.

A subset of climate models add societal factors to a simple physical climate model. These models simulate how population, economic growth, and energy use affect – and interact with – the physical climate. With this information, scientists can produce scenarios of how greenhouse gas emissions may vary in the future. Scientists can then run these scenarios through physical climate models to generate climate change projections.

Climate models include different external forcings for their models. For different greenhouse gas inputs four RCPs (Representative Concentration Pathways) are used: "a stringent mitigation scenario (RCP2.6), two intermediate scenarios (RCP4.5 and RCP6.0) and one scenario with very high GHG (greenhouse gas) emissions (RCP8.5)". Models also include changes in the Earth's orbit, historical changes in the Sun's activity, and volcanic forcing. RCPs only look at concentrations of greenhouse gases, factoring out uncertainty as to whether the carbon cycle will continue to remove about half of the carbon dioxide from the atmosphere each year.

The physical realism of models is tested by examining their ability to simulate contemporary or past climates. Past models have underestimated the rate of Arctic shrinkage and underestimated the rate of precipitation increase. Sea level rise since 1990 was underestimated in older models, but now agrees well with observations. The 2017 United States-published National Climate Assessment notes that "climate models may still be underestimating or missing relevant feedback processes".

Effects

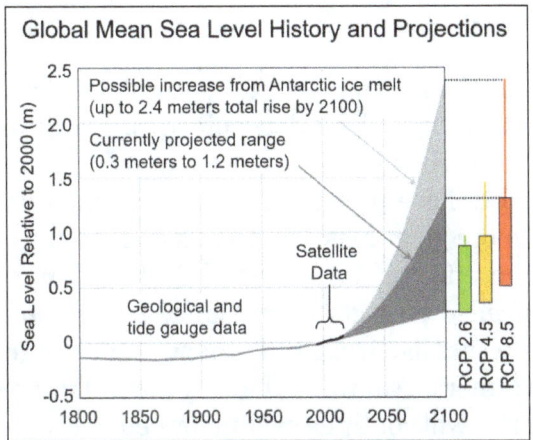

Historical sea level reconstruction and projections up to 2100 published in January 2017 by the U.S. Global Change Research Program for the Fourth National Climate Assessment.

Physical Environment

The environmental effects of global warming are broad and far-reaching. They include effects on the oceans, ice, and weather and may occur gradually or rapidly.

Between 1993 and 2017, the global mean sea level rose on average by 3.1 ± 0.3 mm per year, with an acceleration detected as well. Over the 21st century, the IPCC projects that in a high emissions scenario the sea level could rise by 61–110 cm. The rate of ice loss from glaciers and ice sheets in the Antarctic is a key area of uncertainty since this source could account for 90% of the potential sea level rise. Increased ocean warmth is undermining and threatening to unplug Antarctic glacier outlets, potentially resulting in more rapid sea level rise. The retreat of non-polar glaciers also contributes to sea level rise.

Global warming has led to decades of shrinking and thinning of the Arctic sea ice, making it vulnerable to atmospheric anomalies. Projections of declines in Arctic sea ice vary. While ice-free summers are expected to be rare at 1.5 °C degrees of warming, they are set to occur once every three to ten years at a warming level of 2.0 °C increasing the ice–albedo feedback. Higher atmospheric CO_2 concentrations have led to an increase in dissolved CO_2, which causes ocean acidification. Furthermore, oxygen levels decrease because oxygen is less soluble in warmer water, an effect known as ocean deoxygenation.

Many regions have probably already seen increases in warm spells and heat waves, and it is virtually certain that these changes will continue over the 21st century. Since the 1950s, droughts and heat waves have appeared simultaneously with increasing frequency. Extremely wet or dry events within the monsoon period have increased in India and East Asia. Various mechanisms have been identified that might explain extreme weather in mid-latitudes from the rapidly warming Arctic, such as the jet stream becoming more erratic. The maximum rainfall and wind speed from hurricanes and typhoons are likely increasing.

Long-term effects of global warming: On the timescale of centuries to millennia, the magnitude of global warming will be determined primarily by anthropogenic CO_2 emissions. This is due to carbon dioxide's very long lifetime in the atmosphere. The emissions are estimated to have prolonged the current interglacial period by at least 100,000 years. Because the great mass of glaciers and ice caps depressed the Earth's crust, another long-term effect of ice melt and deglaciation is the gradual rising of landmasses, a process called post-glacial rebound. This could be facilitating seismic and volcanic activity in places like Iceland. Tsunamis could be generated by submarine landslides caused by warmer ocean water thawing ocean-floor permafrost or releasing gas hydrates. Sea level rise will continue over many centuries.

Abrupt climate change, tipping points in the climate system: Climate change could result in global, large-scale changes. Some large-scale changes could occur abruptly, i.e. over a short time period, and might also be irreversible. One potential source of abrupt climate change would be the rapid release of methane and carbon dioxide from permafrost, which would amplify global warming. Another example is the possibility for the Atlantic Meridional Overturning Circulation to slow or shut down. This could trigger cooling in the North Atlantic, Europe, and North America.

Biosphere

In terrestrial ecosystems, the earlier timing of spring events, as well as poleward and upward shifts in plant and animal ranges, have been linked with high confidence to recent warming. It is expected that most ecosystems will be affected by higher atmospheric CO_2 levels and higher global temperatures. Global warming has contributed to the expansion of drier climatic zones, such as, probably, the expansion of deserts in the subtropics. Without substantial actions to reduce the rate of global warming, land-based ecosystems risk major shifts in their composition and structure. Overall, it is expected that climate change will result in the extinction of many species and reduced diversity of ecosystems. Rising temperatures push bees to their physiological limits, and could cause the extinction of bee populations.

The U.S. Geological Survey projects that reduced sea
ice from climate change will lower the population
of polar bears by two-thirds by 2050.

The ocean has heated more slowly than the land, but plants and animals in the ocean have migrated towards the colder poles as fast as or faster than species on land. Just as on land, heat waves in the ocean occur more due to climate change, with harmful effects found on a wide range of organisms such as corals, kelp, and seabirds. Ocean acidification threatens damage to coral reefs, fisheries, protected species, and other natural resources of value to society. Higher oceanic CO_2 may affect the brain and central nervous system of certain fish species, which reduces their ability to hear, smell, and evade predators.

Humans

The effects of climate change on human systems, mostly due to warming and shifts in precipitation, have been detected worldwide. The future social impacts of climate change will be uneven across the world. All regions are at risk of experiencing negative impacts, with low-latitude, less developed areas facing the greatest risk. Global warming has likely already increased global economic inequality, and is projected to do so in the future. Regional impacts of climate change are now observable on all continents and across ocean regions. The Arctic, Africa, small islands, and Asian megadeltas are regions that are likely to be especially affected by future climate change. Many risks increase with higher magnitudes of global warming.

A helicopter drops water on a wildfire in California.
Drought and higher temperatures linked to climate
change are driving a trend towards larger fires.

Food and Water

Crop production will probably be negatively affected in low-latitude countries, while effects at northern latitudes may be positive or negative. Global warming of around 4 °C relative to late 20th century levels could pose a large risk to global and regional food security. The impact of climate change on crop productivity for the four major crops was negative for wheat and maize, and neutral for soy and rice, in the years 1960–2013. Up to an additional 182 million people worldwide, particularly those with lower incomes, are at risk of hunger as a consequence of warming. While increased CO_2 levels help crop growth at lower temperature increases, those crops do become less nutritious. Based on local and indigenous knowledge, climate change is already affecting food security in mountain regions in South America and Asia, and in various drylands, particularly in Africa. Regions dependent on glacier water, regions that are already dry, and small islands are also at increased risk of water stress due to climate change.

Health and Security

Generally, impacts on public health will be more negative than positive. Impacts include the direct effects of extreme weather, leading to injury and loss of life; and indirect effects, such as undernutrition brought on by crop failures. Temperature rise has been connected to increased numbers of suicides. Climate change has been linked to an increase in violent conflict by amplifying poverty and economic shocks, which are well-documented drivers of these conflicts. Links have been made between a wide range of violent behaviour including fist fights, violent crimes, civil unrest, and wars. Climate change may also lead to new human diseases. For example, while ordinary temperatures usually kill off the yeast Candida auris before it infects humans, three strains have recently appeared in widely separate regions, leading researchers to postulate that warmer temperatures are driving it to adapt to higher temperatures at which it can more readily infect humans.

Livelihoods, Industry and Infrastructure

In small islands and mega deltas, inundation from sea level rise is expected to threaten vital infrastructure and human settlements. This could lead to homelessness in countries with low-lying areas such as Bangladesh, as well as statelessness for populations in island nations, such as the Maldives and Tuvalu. Climate change can be an important driver of migration, both within and between countries.

The majority of severe impacts of climate change are expected in sub-Saharan Africa and South-East Asia, where existing poverty is exacerbated. Current inequalities between men and women, between rich and poor and between people of different ethnicity have been observed to worsen as a consequence of climate variability and climate change. Existing stresses include poverty, political conflicts, and ecosystem degradation. Regions may even become uninhabitable, with humidity and temperatures reaching levels too high for humans to survive.

Global Warming Potential

Human activities have led to increases in the concentration of a number of gases which are effective in absorbing and reemitting radiation in the infrared part of the electro-magnetic spectrum, leading to a strengthening of the so-called greenhouse effect. The primary gases responsible for this strengthening are carbon dioxide (CO_2), methane (CH_4), tropospheric ozone (O_3), the chlorofluorocarbons (CFCs) and nitrous oxide (N_2O). Many of the proposed replacements for CFCs, such as the HCFCs and HFCs, are also greenhouse gases. Other gases, such as carbon monoxide (CO) and the nitrogen oxides (NO_x), as well as hydrocarbons (HCs), are not themselves greenhouse gases but influence the concentration of one or more greenhouse gases and thus indirectly affect the strength of the greenhouse effect.

Certain costs will be entailed in reducing the emissions of any one of these gases (whether economic or other kinds of costs). In developing an overall strategy to reduce emissions of greenhouse gases, it is important to be able to quantitatively compare the contribution to the greenhouse effect of a given quantity of each of the gases. The relative unit contribution of each gas to the greenhouse effect can then be compared with the unit cost of emission reduction in instances where there is a trade off between reducing emissions of one gas more and another gas less, or where measures to reduce emissions of one gas might lead to an increase in emissions of another gas.

Unfortunately, the task of quantitatively comparing the greenhouse effectiveness of different gases is fraught with numerous difficulties, both conceptual and analytic. These difficulties arise from the fact that the heat trapping ability of a given gas depends on a number of parameters which change through time, that indirect chemical effects are involved and, most importantly, that the average lifespan in the atmosphere of each gas is different. Thus, if equal quantities of each gas are emitted into the atmosphere at a given time (or conversely, equal emissions are avoided), the relative amounts of the gases remaining in the atmosphere continuously change through time.

Several authors have attempted to overcome the problem of differing lifespans by introducing an index based on the ratio of radiative forcing by different gases integrated over a specific time interval. The expectation is that such indices will be used by policy makers during the negotiation of overall limitations of greenhouse gases, either to permit trading between gases or to allow a given state to choose the specific combination of greenhouse gas emission reductions that it wishes in achieving a given reduction in total greenhouse heating.

The Problem

Given that the relative greenhouse effect of equal emissions of different gases changes over time, one way to quantitatively compare two or more gases is to compare the integrated heating effect over all time. If $C_i(t)$ is the concentration of gas i at time t following the emission of a unit amount at time t = 0, $f_i(t)$ is the heat trapping ability at the same time per unit of concentration, and $C_c(t)$ and $f_c(t)$ the corresponding quantities for CO_2, then the GWP of gas i relative to CO_2 could be given by:

$$GWP = \frac{\int_0^\infty f_i(t)C_i(t)dt}{\int_0^\infty f_c(t)C_c(t)dt}$$

A similar approach is used in calculating the ozone depleting potential (ODP) of different ozone depleting gases: one compares the ozone depletion per unit of concentration of two gases, times the concentration remaining after a pulse input, integrated over all time as the concentrations of the gases continously decrease. For a linear response to CFC emissions, this is mathematically equivalent to considering a continuous and equal rate of emission of two different gases, and comparing the steady state ozone depletion resulting from the two gases.

There is no conceptual problem in making this comparison (whether for ODPs or GWPs) as long as both gases have a finite atmospheric lifetime or, conversely, as long as both gases asymptote to a steady state concentration given a continuous and constant rate of emission. This condition is satisfied for all gases of interest except CO_2. Unlike all other gases emitted in the atmosphere by humans, CO_2 does not have a chemical or photochemical sink within the atmosphere itself. Removal of CO_2 is therefore dependent on exchanges with other carbon reservoirs. Following the emission of a hypothetical pulse of CO_2 into the atmosphere, about 10% will be removed within two years through gaseous diffusion into the mixed layer of the ocean. Subsequent removal is dependent on downward mixing into the deeper ocean. Ultimately, about 85% of the initial pulse can be removed by mixing into the ocean, but only over a time period of several hundred to a thousand years. Another 10% of the initial pulse will be removed by dissolution of marine carbonate sediments, but over a period of several thousand years. The remaining 5% or so is removed over a period of 100 000 years through silicate rock weathering.

Three-dimensional models of ocean chemistry, circulation and deep mixing have been used to determine the rate of decrease in concentration C of a pulse of CO_2, neglecting the geological timescale processes of carbonate sediment dissolution and silicate rock weathering. The concentration of remaining CO_2 following a unit pulse emission can be conveniently approximated by:

$$C_c(t) = \sum_{i=0}^4 A_i \exp\left(-\frac{t}{\tau_i}\right)$$

where the values of A_i and τ_i are given in table. Since $\tau_0 = \infty$ c this can be written as:

$$C_c(t) = A_0 + \sum_{i=1}^4 A_i \exp\left(-\frac{t}{\tau_i}\right)$$

As $t \to \infty$, a fraction Ao of the initial impulse remains in the atmosphere, all other terms in Equation above decaying to zero. As previously noted, this CO_2 will eventually be removed by carbonate dissolution and silicate rock weathering, but the rate of removal is so slow that it is zero for all practical purposes. Thus, for any constant non-negligible anthropogenic emission rate, the CO_2 concentration does not asymptote to a steady state value, but increases indefinitely. Hence, the lower integral in Equation is infinite.

Table: Values of the constants A_i and τ_i in Equation, as given in Maier-Reimer and Hasselmann.

i	A_i	τ_i
0	0.131	∞
1	0.201	362.9
2	0.321	73.6
3	0.249	17.3
4	0.098	1.9

This problem has been resolved in two ways. One is to set $T_0 < \infty$, which is equivalent to assuming that the asymptotic airborne fraction Ao decreases through time. Lashof and Ajuha, for example, set A_0 = 1000 years. However, at the century timescale of human interest, A_0 will actually increase with time because of the increase in the oceanic buffer factor, so that the ability of the ocean to absorb CO_2 inputs to the atmosphere decreases. Only at the timescale of several oceanic overturnings, of the order of several thousand years, will A_0 decrease, as a result of dissolution of carbonate sediments. Resolving the problem of infinite integrals by letting Ao decay to zero with the relatively short time constant of a thousand years serves to underestimate the longterm importance of CO_2 emissions relative to emissions of other greenhouse gases.

An alternative solution to this problem is to retain $\tau_0 = \infty$ but to integrate the radiative forcing arising from an impulse input only up to a time horizon T. Thus:

$$GWP = \frac{\int_0^T f_i(t)C_i(t)dt}{\int_0^T f_c(t)C_c(t)dt}$$

This approach has been adopted by the Intergovernmental Panel on Climatic Change (IPCC) and in subsequent work by Lashof and coworkers published by the Advisory Group on Greenhouse Gases (AGGG). The GWP so defined depends on the choice of time horizon: for gases having a shorter atmospheric residence time than the average for CO_2, GWP decreases as longer time horizons are considered.

The foregoing outlines the conceptual problem in determining GWPs for different gases. In the approaches used so far, the conceptual problem reduces to deciding on a choice of time horizon. Short time horizons are appropriate if one is more concerned with potential climatic changes during the next few decades or if one is concerned with the rate of climatic change, where the rapid build-up of short lived greenhouse gases would tend to provoke initial rapid rates of warming. Longer time horizons are more appropriate if one is more concerned about long-term, chronic effects of climatic warming (such as sea level rise). There are, however, a whole series of uncertainties dealing with the hard science - in particular, determining the direct and indirect radiative heating effects of different gases (the f_i values above) and the time constants for removal from the atmosphere of CO_2 and other gases (the τ values above). These uncertainties are briefly outlined below.

Table: Radiative heating effect per molecule and per unit mass of various greenhouse gases relative to CO_2.

Trace gas	Heating relative to CO_2 per molecule	Per unit mass
CO_2	1	1
CH_4	26	72
N_2O	206	206
CFC-11	12400	3970
CFC-12	15800	5750
HCFC-22	10700	5440
CF_3Br	16000	4730

Uncertainties in Calculation

Radiative Effects

The globally and annually averaged trapping of infrared radiation at the tropopause (the base of the stratosphere) provides a first order estimate of the overall warming effect of a given greenhouse gas. This trapping involves two components: a decrease in upward radiation from the troposphere (which underlies the stratosphere) and an increase in downward radiation from the stratosphere. The net effect on infrared radiation depends on:

- Absorption and emission coefficients as a function of gas amount, pressure, temperature, and wavelength, which can be determined to high accuracy from laboratory spectroscopic measurements.

- The degree of overlap in the absorption of the gas in question with absorption by other gases, and hence on the concentrations of all other gases.

- The atmospheric temperature profile and cloudiness.

The absorption coefficients of many gases do not vary linearly with gas concentration: the radiative heating of a unit of a given gas thus depends on the preexisting concentration of the gas and thus changes through time. It also depends on the concentrations of other gases which absorb in the same spectral regions. For these reasons, the heat trapping ability f_i depends on the scenarios of future concentrations of most or all of the greenhouse gases. All calculations of GWP to date have assumed time invariant f_i values, appropriate to present day conditions.

Table: Compares the average radiative heating per molecule of various gases relative to CO_2, as determined by the IPCC except for CH_4, which is based on Lelieveld and Crutzen. The values given in table assume the gases to be uniformly mixed in the atmosphere. This is a valid assumption for the gases shown in table, but not for O_3, whose concentration varies substantially in all three dimensions. The effectiveness of a given tropospheric gas in reducing upward infrared radiation at the tropopause is greater the colder it is relative to the surface; removal of an ozone molecule in the upper troposphere (which is coldest relative to the surface) causes a much greater decrease in the greenhouse effect than removal of an ozone molecule near the surface. In the case of stratospheric ozone, an ozone decrease in the lower stratosphere is more effective in reducing downward radiation at the tropopause than a decrease in the middle or upper stratosphere. Offsetting this decrease to varying degrees is an increase in downward solar radiation at the tropopause. Hence, the net effect of ozone losses depends critically on the vertical distribution of changes, and can be one of heating or cooling.

Atmospheric Lifetime

CO_2: Observations from polar ice cores indicate that the atmospheric CO_2 concentration was close to constant for several hundred years prior to the industrial revolution. Since 1800, the atmospheric CO_2 concentration has increased by 25% and there is no doubt that this increase is due to anthropogenic activities. However, the observed increase - the so-called airborne fraction - is only about 50% of the cumulative emissions since 1800. The remaining 50% has already been absorbed by some combination of the oceans and terrestrial biosphere. Unfortunately, we are not able to determine the relative importance of these two sinks. Some evidence indicates that the oceans could not have absorbed more than 1/3 of the non-airborne CO_2 and that the northern temperate latitude biosphere is a major carbon sink. This, however, is contradicted by isotopic a3C data, which indicate that the northern high latitude oceans are a larger sink than the northern biosphere.

Determination of the current CO_2 sinks, and how these sinks are likely to change with increasing CO_2 emissions, is crucial to the calculation of GWPs. If the primary sink is the terrestrial biosphere then continuing forest destruction, nutrient limitations or temperature induced increases of respiration could dramatically increase the lifetime of CO_2 in the atmosphere. If the primary sink is the oceans, then climatic change-induced changes in ocean circulation or biological productivity could atmospheric CO_2 lifetime. Sarmiento and Toggweiler, and Baes and Killough concluded that decreased oceanic overturning would reduce atmospheric CO_2, while Baes, and Broecker and Takahashi suggest the opposite. Uncertainty in the removal rates of CO_2 from the atmosphere is a major source of uncertainty in the calculation of GWPs over a time horizon of the next few decades.

CH_4: The primary sink of methane is reaction with atmospheric OH, with soils serving as an additional small sink. The average lifetime of a methane molecule is 8-12 years, based on the global average OH concentration, which is computed from the distribution and concentration trend for methyl chloroform. However, the exact value is sensitive to the atmospheric OH distribution. It is assumed that CH_3CCl_3 is destroyed only by reaction with OH, but recent work suggests that some CH_3C-Cl_3 is taken up by the ocean, is such additional loss could mean that the inferred OH concentration is 5-20% too large, and hence that the lifetime of molecules lost solely by reaction with OH is 5-20% longer than currently estimated. In addition, a recent revision of the CH_4-OH rate constant suggests a further 25% increase in the estimated lifetime of CH_4.

The OH concentration itself depends in complex ways on emissions of CH_4, thus introducing a feedback between CH_4 concentration and its GWP. Uncertainties in the CH_4-OH coupling introduce a further uncertainty in the CH_4 GWP of about 20-30% over a 50 year time horizon. Other chemical effects on the CH_4 GWP are discussed below.

N_2O: The major sink of N_2O is photochemical destruction in the stratosphere, and its estimated lifetime has recently been revised from 150 years to 110 years. The main sources appear to be the oceans, denitrification in aerobic soils, and combustion and biomass burning. However, we cannot at present balance the N_2O budget, which introduces uncertainties in determining its present atmospheric lifetime. As with CO_2, it is possible that changes in ocean circulation or biological productivity as climate changes could significantly alter its atmospheric lifetime.

Halocarbons: Although the most ozone damaging chemicals will be strongly controlled over the coming years, many of the proposed substitutes are important greenhouse gases. Most of these are hydrochlorofluorocarbons (HCFCs), and are destroyed principally by reaction with tropospheric OH. Their lifetime will therefore change in response to changing OH concentration. Additionally, the role of the oceans in absorbing HCFCs is not known.

Effects of Atmospheric Chemistry

Atmospheric chemistry can influence the GWP of a given gas by altering its atmospheric lifetime, primarily through changing OH concentrations. Atmospheric chemistry can also alter the concentration or occurrence of greenhouse gases other than the one being emitted. Such indirect effects should be added to the direct heating effect of a given gas in assessing its GWP. The major indirect effects are discussed below.

CO_2: An increase of CO_2, unlike that of other greenhouse gases, tends to cool the stratosphere. To the extent that this favours the formation of polar stratospheric clouds, this will worsen stratospheric ozone depletion. Such ozone depletion will have a cooling effect, although further effects through negative impacts of increased UV radiation on terrestrial and oceanic photosynthesis could lead to increased CO_2 fluxes to the atmosphere. The radiative impact of both effects is likely to be small compared to uncertainties concerning the atmospheric lifetime of CO_2, although the implications for stratospheric ozone depletion are serious as CO_2 induced stratospheric cooling would reinforce stratospheric cooling due to ozone depletion itself.

CH_4: Methane can indirectly affect the trapping of infrared radiation in the following ways:

- Oxidation of methane in the stratosphere produces stratospheric water vapour, which is not subject to the rapid removal of the lower atmosphere (where the mean lifespan is eight days), thereby increasing greenhouse heating by as much as 1/3 of the direct effect of methane according to the IPCC but by only 7% according to Lelieveld and Crutzen.

- The afore mentioned stratospheric water vapour production leads to ozone destruction in the upper stratosphere.

- Methane tends to increase stratospheric ozone in the lower stratosphere by forming HC1 from free CI, thereby removing CI from the catalytic ozone destruction cycle.

- Methane contributes to the formation of tropospheric ozone under conditions of high NO_x; the net heating effect of all ozone change is estimated to be 40-60% by Rotmans et al and 20--40% by Lelieveld and Crutzen, while Owens et al estimate the net effect of combined water vapour and ozone changes to be 76% of the direct effect.

- By altering the OH concentration, methane emissions alter the atmospheric lifespan of other greenhouse gases besides its own; for some scenarios, up to a 50% increase in the lifetime of methane and CFC substitutes is estimated, with heating effects of 20-50% of the direct effect.

- Methane is ultimately oxidized to CO_2.

Both the direct and indirect methane GWPs decrease as the time horizon increases, although the indirect GWP as a percentage of the direct GWP can increase or decrease through time, depending on the gas emission scenario. For constant emissions of all gases, Lelieveld and Crutzen find that the indirect methane GWP adds 38% to the direct GWP at a 20 year time horizon but only 25% at a 100 year time horizon, while for a scenario of increasing gas emissions the indirect effects add 60% and 74% to the direct effects at 20 and 100 year time horizons respectively. The impact of CH_4 (and CO) emissions on OH and tropospheric ozone depends on the background NO_x concentration, which is highly variable. The mean atmospheric lifetime of methane is long (8-12 years) compared to the time to mix throughout the atmosphere (months within hemispheres, one year for mixing between hemispheres), so that the net effect of a given methane emission on OH and ozone is largely independent of where the emission occurs, although computation of the globally averaged effect requires integration over regions of differing background chemistry and radiative fluxes.

In light of these uncertainties, the IPCC update declined to estimate an indirect methane GWP, while pointing out that the indirect GWP could be comparable in magnitude to the direct GWP.

N_2O: N_2O is an important source of reactive nitrogen in the stratosphere; it contributes to ozone destruction both through gaseous phase chemistry and through the formation of polar stratospheric clouds, which are implicated in the Antarctic ozone hole. [31] These effects probably reduce the GWP of N_2O but are also probably small.

CFCs: To the extent that CFC emissions alone are responsible for stratospheric ozone depletion (ie if CO_2 induced stratospheric cooling were negligible or had a negligible effect), the net warming effect of CFC emissions is reduced. A recent analysis for the period 1970--81 indicates that the stratospheric ozone depletion in northern hemisphere mid latitudes during this period masked half of the incremental heating due to increase of CO_2 during the same period. Limited data from a single southern hemisphere mid-latitude location also imply a surface cooling due to ozone changes, while indirect evidence suggests that ozone changes at low latitudes have also had a net cooling effect. Since the direct heating effect of CFC increases between 1970 and 1980 was only 40% that due to the CO_2 increase, the net effect of CFCs could be one of cooling. The net effect of ozone changes depends critically on the vertical distribution of changes, as ozone losses above 30 km have a net warming effect due to the fact that the increased penetration of solar radiation more than compensates the reduction in infrared emission to the troposphere. Further complicating the picture is the fact that stratospheric ozone loss allows an increase in penetration of ultraviolet radiation to the troposphere, leading to increased tropospheric ozone and production of OH, thereby tending to reduce the mean atmospheric methane lifespan. There are, therefore, large uncertainties in the determination of the net radiative effect of ozone changes and hence in the net effect of CFCs.

CO: CO emissions, like those of methane, affect tropospheric OH and ozone concentrations and thus indirectly affect greenhouse heating. Three-dimensional model simulations suggest that anthropogenic CO emissions have decreased atmospheric OH at low latitudes and in the southern hemisphere, but have increased atmospheric OH north of 20 ° N. The mean atmospheric lifetime of CO is about two months, which is comparable to atmospheric mixing times, so that the average global effect of CO emissions depends on where the emissions occur. As with methane, CO is ultimately oxidized to CO_2.

NO_x: NO_x emissions contribute to tropospheric ozone formation and hence indirectly add to the greenhouse effect, but also tend to increase atmospheric OH, which will tend to shorten the lifetime of CH_4 and other greenhouse gases, thereby reducing the greenhouse effect. The extent and even occurrence of ozone formation associated with NO_x emissions is highly dependent on the regional atmospheric chemistry, so that, like CO, there is no single GWP for NO_x which can be applied everywhere. Furthermore, NO_x emissions in the upper troposphere (from aircraft) are about 20 times more effective in producing O_3 than are surface emissions, and upper tropospheric ozone is about 1.3 times.

Table: Atmospheric lifespans and global warming potentials per unit mass as computed by the IPCC update for time horizons of 20, 100 and 500 years.

Gas	Average lifespan (years)	20 years	Time horizon 100 years	500 years
CH_4	10.5	35	11	4
N_2O	132	260	270	170
CFC-11	55	4500	3400	1400
CFC-12	116	7100	7100	4300
HCFC-22	15.58	4200	1600	540

More effective, on a molecule-per-molecule basis, in trapping infrared radiation than surface ozone. Consequently, NO_x emissions from high flying aircraft are calculated to have about 30 times the warming effect of equal emissions from surface sources.

Direct GWPs

Table gives GWPs for CH_4, N_2O, CFC-11, CFC-12 and HCFC-22 as computed by the IPCC update; these GWPs include only direct radiative effects. A revision in the estimated lifespan of CO_2, or climate-carbon cycle feedbacks, could dramatically change all of the GWPs given in table.

Potential Applications of GWPs

Given the enormous uncertainties in the calculation of GWPs and their dependence on future scenarios, it is difficult to see how they could be used in any rigorous way for policy analysis. The US administration advocates a basket approach to any greenhouse gas emission reductions, whereby any nation can reduce the mix of greenhouse gases giving an agreed net effect. This requires the ability to quantitatively compare different greenhouse gases. As indicated above, we are far from the point where reliable intercomparisons can be made for a given time horizon. Indeed, it is extremely unlikely that we will be able to predict all of the changes affecting the atmospheric lifetime of CO_2 and other greenhouse gases; such changes as do occur are likely to occur in a non-uniform manner, implying strong nonlinearities in the response to successive emission increments. Oceanic circulation, changes in which could significantly affect the mean atmospheric CO_2 lifetime, can change dramatically following subtle changes in precipitation and evaporation fields. In addition, the problem of which time horizon to use remains and is not subject to scientific determination. A further problem with this potential application of GWPs is the great difficulty in quantifying and monitoring the sources and sinks of most gases, as discussed by Victor.

Another potential application of GWPs is in assessing the net effect of measures which reduce emissions of one greenhouse gas but increase emissions of another gas. For example, switching from coal to natural gas for electricity generation might lead to an increase of methane emissions (depending on the relative magnitude of methane leakage from natural gas distribution and methane seepage from coal mining) while significantly reducing CO_2 emissions. This potential application suggests an alternative GWP index tied to the lifetime of the particular investment decision under consideration.

An Alternative GWP Index

The change in concentration $y(t)$ of a given gas following a continuous emission at a rate $x(t)$ is given by the convolution of the emission rate with the impulse response, $C(t)$. That is:

$$y(t) = \int_0^t x(t')C(t-t')dt'$$

Equation assumes a linear relationship between emission and concentration, so that successive emission pulses have the same incremental effect on concentration. For the case of CH_4, $C(t)$ can be represented by a simple exponential decay, $\exp(-t/\tau)$, so that Equation yields:

$$y_{CH4}(t) = \tau x_{CH4}\left(1-\exp\left(-\frac{t}{\tau}\right)\right)$$

for a constant emission rate x_{CH4}. As $t \to \infty$, the methane concentration approaches the steady state value τx_{CH4}. For the case of CO_2, $C(t)$ can be given by Equation, so that the CO_2 concentration change is:

$$y_{CO2}(t) = x_{CO2}\left(A_0 t + \sum_i \tau_i \left(1-\exp\left(-\frac{t}{\tau_i}\right)\right)\right)$$

where x_{CO2} is a constant CO_2 emission rate. If the radiative forcing per molecule is constant in time, the GWP as defined by Equation is mathematically equivalent to comparing the radiative heating of two gases at time T after a continuous and equal emission of both gases up to time T. Thus, for CH_4 the GWP as given by Equation is equivalent, for these conditions, to:

$$\frac{y_{CH4}(T)}{y_{CO2}(T)} \frac{f_{CH4}}{f_{CO2}}$$

where f_{CH4} and f_{CO2} are the time invariant radiative forcings per unit concentration.

In the more general case where the radiative forcing per molecule varies with time, the GWP is given by:

$$GWP(t) = \frac{\int_0^T f_i(t)C_i(T-t')dt'}{\int_0^T f_c(t)C_c(T-t')dt'}$$

where $f_i(t)$ and $f_c(t)$ are the time dependent radiative forcings for gas i and CO_2, and $x_i(t) = x_c(t) = x_i$ and thus cancel from the numerator and denominator of the above expression.

Casting the GWP in the form given by Equation represents an important conceptual shift from that of $GWP = \dfrac{\int_0^T f_i(t)C_i(t)dt}{\int_0^T f_c(t)C_c(t)dt}$. Rather than viewing the GWP as the ratio of the integrated radiative heating due to a single pulse emission of both gases at time $t = 0$, it can be viewed as being approximately equal to the ratio of instantaneous radiative forcings at the end of a given time period, assuming equal and continuous emissions during the entire time interval of interest.

In the case of investment decisions involving emissions of two or more gases over a finite period of time, the time horizon T should be chosen equal to the lifespan of the end-use technology. If one is interested in the relative greenhouse impact of switching from oil to natural gas for automobiles, home heating, or electricity generation, taking into account the greenhouse forcing of CH_4, then $T = 10, 20$ and 40 years respectively are reasonable choices.

If one is concerned with the greenhouse implications of a given investment decision beyond the lifetime T of the decision, then it is appropriate to assume that the emission rate $x(t) = 0$ for $t > T$. That is, in assessing the relative greenhouse impact of different investment options, one should assume that emissions of all the associated greenhouse gases occur only for as long as the lifetime of the investment decision, even if one is interested in longer time horizons. If it is decided to replace an old fossil fuel using end-use technology at the end of its economically useful life with a new fossil fuel using technology, that represents a separate investment decision, the impact of which should not be incorporated in analysis of the greenhouse implications of the first investment decision. Furthermore, it is possible that a fossil fuel technology installed now might be replaced with a non-fossil fuel technology at the end of its life.

Based on the above discussion, the GWP for times $t \geq T$ is given by:

$$GWP(t) = \frac{\int_0^T f_i(t)C_i(t-t')dt'}{\int_0^T f_c(t)C_c(t-t')dt'}, t \geq T \text{ of which } GWP(t) = \frac{\int_0^T f_i(t)C_i(T-t')dt'}{\int_0^T f_c(t)C_c(T-t')dt'} \text{ is a special}$$

case.

Table: Global warming potentials per unit mass for CH_4 and HCFC-22 for different time horizons and investment lifetimes as computed using Equation.

	Time horizon years 20		40			100				500			
Investment lifetime (years)	10	20	10	20	40	10	20	40	100	10	20	40	500
CH_4	26.7	42.7	6.0	9.3	28.3	1.1	1.1	1.2	15.3	1.0	1.0	1.0	6.1
HCFC-22	3093	4036	1104	1509	2949	36	51	110	1628	0.0	0.0	0.0	580

For purposes of illustrating the difference in GWP for time horizons t beyond the lifetime T of an investment decision, assume that $f_i(t)$ and $f_c(t)$ are constant. We shall consider the direct GWP of methane and HCFC-22 using the relative radiative heatings and lifetimes given in tables respectively, and taking into account the CO_2 produced from oxidation of methane. The GWP for methane, assuming that emissions occur only during the lifetime T of an investment decision, is given by:

$$GWP(t) = \frac{f_{CH4} \int_0^T C_{CH4}(t-t')x_0 dt' + f_C \int_0^t C_C(t-t')x_c(t')dt'}{f_C \int_0^T C_C(t-t')x_0 dt'}$$

where the CO_2 source $x_c(t)$ due to methane oxidation is given by:

$$x_c(t) = \begin{cases} x_0\left(1-\exp)\left(-\frac{t}{\tau}\right)\right) & t \leq T \\ x_0 \exp\left(-\frac{t}{\tau}\right)\left(\exp\left(\frac{T}{\tau}\right)-1\right)_{t>T} \end{cases}$$

Table gives the GWP, as computed by Equation, for time horizons t of 10, 20, 40, 100 and 500 years. The IPCC/AGGG method implicitly assumes an investment lifetime equal to the time horizon, while for the new method, results are given assuming investment lifetimes T of 10, 20 and 40 years.

For time horizons equal to the lifetime of the particular investment decision, the GWP index proposed here is equivalent to that of the IPCC. The GWP values given in table for this case differ from the IPCC values given in table because of the use of a different model for absorption of CO_2 by the oceans and because of the inclusion here of the CO_2 oxidation product in the case of methane. For time horizons beyond the investment lifetime, the GWPs rapidly fall to one in the case of CH_4, or zero in the case of HCFC-22 (indicating in both cases, that only CO_2 is important), whereas they remain large using the IPCC/AGGG approach. The GWPs rapidly approach one or zero using the new approach because of the short atmospheric lifespan of CH_4 and HCFC-22 relative to CO_2 and relative to the time since the cessation of emissions. For example, the average lifespan of CH_4 is 10 years, compared to removal time constants of up to 363 years for CO_2. A 100 year time horizon is six mean lifespans beyond the last emission of CH_4 for a 40 year investment life, so that there is negligible CH_4 remaining. Since the cumulative carbon emission is the same for CH_4 and CO_2 emissions, and all the CH_4 is ultimately converted to CO_2, the GWP rapidly asymptotes to a value of 1.0.

An interesting property of the GWP index proposed here is that, for gases whose atmospheric lifetime is short compared to that of CO_2, it is less sensitive to changes in the relative radiative forcing per molecule or greenhouse gas lifetimes for time horizons longer than the investment lifetime. Given the large uncertainties in these parameters, this robustness is important from a policy point of view.

Practical Implications of GWPs

It is argued above that the scientific uncertainties in the calculation of GWPs, coupled with the great difficulties in measuring anthropogenic emissions of most gases, preclude the use of GWPs as part of an international agreement involving overall greenhouse forcing limits or trading between gases. The only foreseeable application of GWP is in determining the net effect of fuel switching in order to screen potential greenhouse gas emission reduction options. In this case, the great uncertainty associated with GWPs for time horizons comparable to the investment lifetime can be accounted for by allowing only those fuel switching options which are calculated to reduce total greenhouse forcing by a large fraction. However, when one screens fuel switching options in terms of cost-effectiveness as part of an integrated greenhouse gas emission reduction strategy, one finds that the net greenhouse reduction of many options which pass the economic test is so large that even the most extreme GWP values do not cause the fuel switching option to be rejected.

The effective CO_2 emission factor for a mix of greenhouse gases is given by:

$$F_{eff} = E(F_{CO2} + \sum F_i GWP_i)$$

where E is the fossil fuel consumption (G J) at the point of use, F_{CO2} is the CO_2 emission factor (kg/GJ), and Fi and GWPi are the emission factors (kg/GJ) and GWPs for the other emitted gases. The CO_2 emission factors for coal, oil, and natural gas are 88-95 kg/GJ, 68-73 kg/GJ and 49.5 kg/GJ respectively. In comparing different fuels, account also must be taken of the emissions of all gases associated with the extraction, processing, and transportation of the fuel to the point of use. This requires multiplying the above CO_2 emission factors by average markup factors of 1.04, 1.13 and 1.18 for coal, oil and natural gas respectively. 41 Methane emission factors range from 0.019 kg/GJ for typical lignites to 0.554 kg/GJ for typical bituminous coals (based on Barnes and Edmonds, assuming heating values of 7000 Btu/lb for lignite and 12000 Btu/lb for bituminous coal), while each percentage leakage of natural gas corresponds to an emission factor of 0.182 kg/GJ (based on the higher heating value for methane of 55 MJ/kg and assuming natural gas to be 100% methane). Examples of fuel switching with large effective CO_2, emission reductions are given below.

Coal to Natural Gas for Electricity Generation

If natural gas combined cycle (46-48% efficiency) or cogeneration (65-95% marginal efficiency) replaces conventional coal-fired electricity generation (33% efficiency), CO_2 emission per kilowatt hour of electricity is reduced by a factor of two to four. Because almost two to three times more coal primary energy is used than natural gas primary energy in this comparison, methane emissions will also be reduced if the methane emission factor for natural gas is no more than two to three times larger than for coal. Natural gas emission factors in Western countries are undoubtedly smaller than for bituminous coal, which is the coal most often used for electricity generation.

Electric Resistance to High Efficiency Natural Gas for Heating

Switching from electric resistance heating to high efficiency (92%) natural gas heating reduces CO_2 emissions by almost a factor of four if the electricity is coal fired. This reduction is so large, and the reduction in primary energy use so large, that a significant net greenhouse emission reduction occurs even for natural gas leakage rates as large as 1-2%.

Mid-efficiency Oil to High Efficiency Natural Gas for Heating

Switching from a mid-efficiency (78%) oil furnace to a high efficiency natural gas furnace will reduce CO_2 emissions by about 40%. This is a large enough reduction to give a significant net benefit after allowing for limited (no more than 1%) natural gas leakage and an extreme (twice the direct) GWP for methane.

Electric chillers to advanced natural gas absorption chillers: An alternative to electric chillers (coefficient of performance (COP) = 3-4) is an advanced natural gas absorption chiller (COP = 2-2.5). Although the electric chiller has a higher COP, greater primary energy is required per unit of cooling if the electricity is derived from a conventional coal fired power plant. The CO_2 emission reduction for this switch ranges from 64% to 78%. If natural gas absorption chillers displace electric chillers powered by electricity from cogeneration, on the other hand, there may be no CO_2 emission reduction. However, absorption chillers do not require CFCs or chlorine containing CFC substitutes; although the GWP for these gases might very well be negative (if the effect of ozone loss is included), options which do not require chlorine containing substitutes are likely to be increasingly favoured to provide greater protection to the ozone layer (many proposed CFC substitutes have a significant ODP on a 10--20 year time horizon, with small ODPs only on much longer time horizons).

It should be noted that even in cases where fuel switching for heating and cooling would increase methane emissions, it is still possible to achieve simultaneous reductions in

CO_2 and methane if fuel switching is combined with other measures such as thermal envelope improvements and reduction of internal cooling loads through adoption of more efficient lighting and machines. Such 'bundling' of measures is attractive purely as a CO_2 emission reduction measure because savings in downsizing of heating and cooling equipment when they are due for replacement can offset part of the cost of envelope or lighting improvements. Thus, in most cases involving fuel switching it is possible to achieve simultaneous reductions in both CO_2 and methane, thus rendering the GWP index irrelevant to the decision of whether or not to switch.

RADIATIVE FORCING

Radiative forcing or climate forcing is the difference between insolation (sunlight) absorbed by the Earth and energy radiated back to space. The influences that cause changes to the Earth's climate system altering Earth's radiative equilibrium, forcing temperatures to rise or fall, are called climate forcings. Positive radiative forcing means Earth receives more incoming energy from sunlight than it radiates to space. This net gain of energy will cause warming. Conversely, negative radiative forcing means that Earth loses more energy to space than it receives from the sun, which produces cooling.

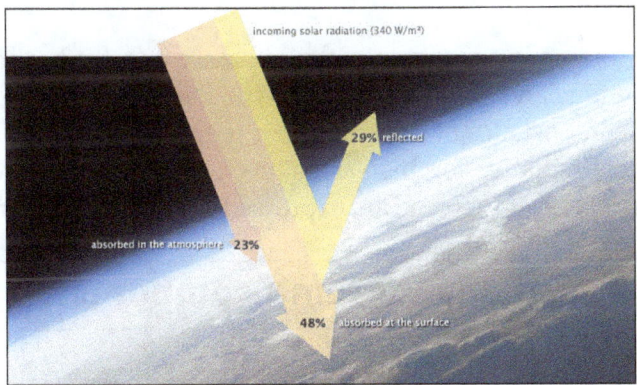

Incoming solar radiation.

Typically, radiative forcing is quantified at the tropopause or at the top of the atmosphere (often accounting for rapid adjustments in temperature) in units of watts per square meter of the Earth's surface. Positive forcing (incoming energy exceeding outgoing energy) warms the system, while negative forcing (outgoing energy exceeding incoming energy) cools it. Causes of radiative forcing include changes in insolation and the concentrations of radiatively active gases, commonly known as greenhouse gases, and aerosols.

Radiation Balance

Atmospheric gases only absorb some wavelengths of energy but are transparent to others. The absorption patterns of water vapor (blue peaks) and carbon dioxide (pink peaks) overlap in some wavelengths. Carbon dioxide is not as strong a greenhouse gas as water vapor, but it absorbs energy in wavelengths (12-15 micrometers) that water vapor does not, partially closing the "window" through which heat radiated by the surface would normally escape to space.

Almost all of the energy that affects Earth's climate is received as radiant energy from the Sun. The planet and its atmosphere absorb and reflect some of the energy, while long-wave energy is radiated back into space. The balance between absorbed and radiated energy determines the average global temperature. Because the atmosphere absorbs some of the re-radiated long-wave energy, the planet is warmer than it would be in the absence of the atmosphere:

The radiation balance is altered by such factors as the intensity of solar energy, reflectivity of clouds or gases, absorption by various greenhouse gases or surfaces and heat emission by various materials. Any such alteration is a radiative forcing, and changes the balance. This happens continuously as sunlight hits the surface, clouds and aerosols form, the concentrations of atmospheric gases vary and seasons alter the groundcover.

IPCC Usage

The Intergovernmental Panel on Climate Change (IPCC) AR4 report defines radiative forcings as:

> "Radiative forcing is a measure of the influence a factor has in altering the balance of incoming and outgoing energy in the Earth-atmosphere system and is an index of the importance of the factor as a potential climate change mechanism. In this report radiative forcing values are for changes relative to preindustrial conditions defined at 1750 and are expressed in Watts per square meter (W/m²)."

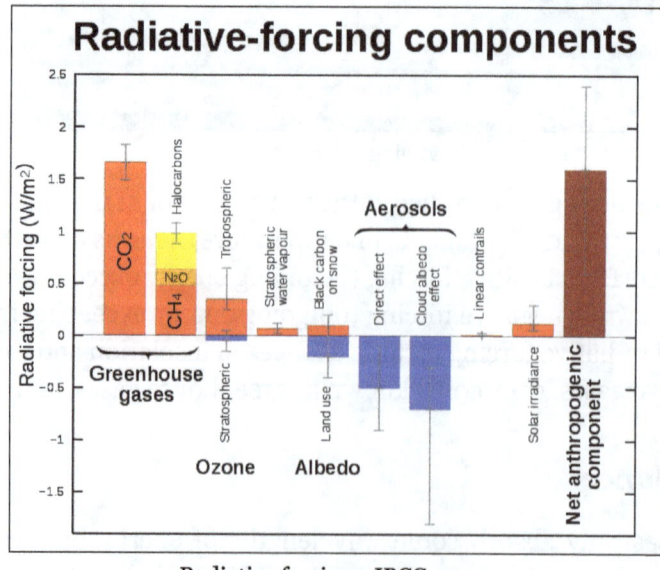

Radiative forcings, IPCC 2007.

In simple terms, radiative forcing is "the rate of energy change per unit area of the globe as measured at the top of the atmosphere." In the context of climate change, the term "forcing" is restricted to changes in the radiation balance of the surface-troposphere system imposed by external factors, with no changes in stratospheric dynamics, no surface and tropospheric feedbacks in operation (i.e., no secondary effects induced because of changes in tropospheric motions or its thermodynamic state), and no dynamically induced changes in the amount and distribution of atmospheric water (vapour, liquid, and solid forms).

Climate Sensitivity

Radiative forcing can be used to estimate a subsequent change in steady-state (often denoted "equilibrium") surface temperature (ΔT_s) arising from that forcing via the equation:

$$\Delta T_s = \lambda \, \Delta F$$

where λ is commonly denoted the climate sensitivity parameter, usually with units K/(W/m²), and ΔF is the radiative forcing in W/m². A typical value of λ, 0.8 K/(W/m²), gives an increase in global temperature of about 1.6 K above the 1750 reference temperature due to the increase in CO_2 over that time (278 to 405 ppm, for a forcing of 2.0 W/m²), and predicts a further warming of 1.4 K above present temperatures if the CO_2 mixing ratio in the atmosphere were to become double its pre-industrial value; both of these calculations assume no other forcings.

Sample Calculations

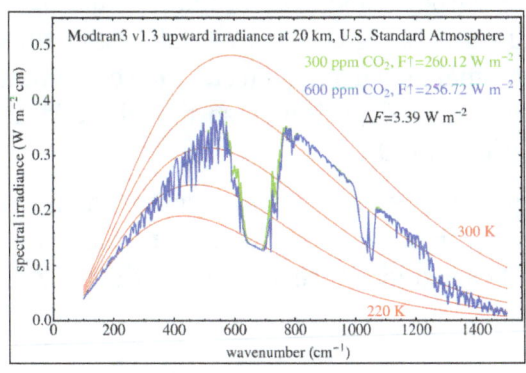

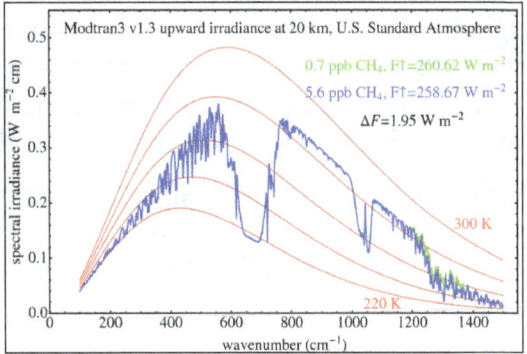

Radiative forcing for doubling CO_2, as calculated by radiative transfer code Modtran. Red lines are Planck curves.

Radiative forcing for eight times increase of CH_4, as calculated by radiative transfer code Modtran.

Solar Forcing

Radiative forcing (measured in watts per square meter) can be estimated in different ways for different components. For solar irradiance (i.e. "solar forcing"), the radiative forcing is simply the change in the average amount of solar energy absorbed per square meter of the Earth's area. Approximating the Earth as a sphere, the Earth's cross-sectional area exposed to the Sun (πr^2) is equal to 1/4 of the surface area of the Earth ($4\pi r^2$), the solar input per unit area is one quarter the change in solar intensity. Since some radiation is reflected, this must be multiplied by the fraction of incident sunlight that is absorbed, $F = (1-R)$, where R is the reflectivity (albedo) of the Earth —approximately 0.3, so F is approximately equal to 0.7. Thus, the solar forcing is the change in the solar intensity divided by 4 and multiplied by 0.7.

Likewise, a change in albedo will produce a solar forcing equal to the change in albedo divided by 4 multiplied by the solar constant.

Forcing due to Atmospheric Gas

For a greenhouse gas, such as carbon dioxide, radiative transfer codes that examine each spectral line for atmospheric conditions can be used to calculate the change ΔF as a function of changing concentration. These calculations might be simplified into an algebraic formulation that is specific to that gas.

For instance, a proposed simplified first-order approximation expression for carbon dioxide would be:

$$\Delta F = 5.35 \times \ln\frac{C}{C_0} \, \mathrm{W\,m^{-2}}$$

where C is the CO_2 concentration in parts per million by volume and C_0 is the reference concentration. The relationship between carbon dioxide and radiative forcing is logarithmic, at concentrations up to around eight times the current value, and thus increased concentrations have a progressively smaller warming effect. Some claim that at higher concentrations, however, it becomes supra-logarithmic so that there is no saturation in the absorption of infrared radiation by CO_2.

A different formula might apply for other greenhouse gases such as methane and N$_2$O (square-root dependence) or CFCs (linear), with coefficients that may be found e.g. in the IPCC reports. While recently a suggests a significant revision of methane IPCC formula.

Related Measures

Radiative forcing is a useful way to compare different causes of perturbations in a climate system. Other possible tools can be constructed for the same purpose: for example Shine et al. say "recent experiments indicate that for changes in absorbing aerosols and ozone, the predictive ability of radiative forcing is much worse we propose an alternative, the 'adjusted troposphere and stratosphere forcing'. We present GCM calculations showing that it is a significantly more reliable predictor of this GCM's surface temperature change than radiative forcing. It is a candidate to supplement radiative forcing as a metric for comparing different mechanisms". In this quote, GCM stands for "global circulation model", and the word "predictive" does not refer to the ability of GCMs to forecast climate change. Instead, it refers to the ability of the alternative tool proposed by the authors to help explain the system response.

Therefore, the concept of radiative forcing has been evolving from the initial proposal, named nowadays instantaneous radiative forcing (IRF), to other proposals that aims to relate better the radiative imbalance with global warming (global surface mean temperature). In this sense the adjusted radiative forcing, in its different calculation methodologies, estimates the imbalance once the stratosphere temperatures has been modified to achieve a radiative equilibrium in the stratosphere (in the sense of zero radiative heating rates). This new methodology is not estimating any adjustment or feedback that could be produced on the troposphere (in addition to stratospheric temperature adjustments), for that goal another definition, named effective radiative forcing has been introduced. In general the ERF is the recommendation of the CMIP6 radiative forcing analysis although the stratospherically adjusted methodologies are still being applied in those cases where the adjustments and feedbacks on the troposphere are considered not critical, like in the well mixed greenhouse gases and ozone. A methodology named radiative kernel approach allows to estimate the climate feedbacks within an offline calculation based on a linear approximation.

IMPACT ON GLOBAL TEMPERATURES

The world is getting warmer. Whether the cause is human activity or natural variability—and the preponderance of evidence says it's humans—thermometer readings all around the world have risen steadily since the beginning of the Industrial Revolution.

According to temperature analysis conducted by scientists at NASA's Goddard Institute for Space Studies (GISS), the average global temperature on Earth has increased by about 0.8° Celsius (1.4° Fahrenheit) since 1880. Two-thirds of the warming has occurred since 1975, at a rate of roughly 0.15-0.20°C per decade.

The global temperature record represents an average over the entire surface of the planet. The temperatures we experience locally and in short periods can fluctuate significantly due to predictable cyclical events (night and day, summer and winter) and hard-to-predict wind and precipitation patterns. But the global temperature mainly depends on how much energy the planet receives from the Sun and how much it radiates back into space—quantities that change very little. The amount of energy radiated by the Earth depends significantly on the chemical composition of the atmosphere, particularly the amount of heat-trapping greenhouse gases.

A one-degree global change is significant because it takes a vast amount of heat to warm all the oceans, atmosphere, and land by that much. In the past, a one- to two-degree drop was all it took to plunge the Earth into the Little Ice Age. A five-degree drop was enough to bury a large part of North America under a towering mass of ice 20,000 years ago.

The maps below show temperature anomalies, or changes, not absolute temperature. They depict how much various regions of the world have warmed or cooled when compared with a base period of 1951-1980. (The global mean surface air temperature for that period was estimated to be 14°C (57°F), with an uncertainty of several tenths of a degree.) In other words, the maps show how much warmer or colder a region is compared to the norm for that region from 1951-1980.

Global temperature records start around 1880 because observations did not sufficiently cover enough of the planet prior to that time. The period of 1951-1980 was chosen largely because the U.S. National Weather Service uses a three-decade period to define "normal" or average temperature. The GISS temperature analysis effort began around 1980, so the most recent 30 years was 1951-1980. It is also a period when many of today's adults grew up, so it is a common reference that many people can remember.

The line plot shows yearly temperature anomalies from 1880 to 2014 as recorded by NASA, NOAA, the Japan Meteorological Agency, and the Met Office Hadley Centre (United Kingdom). Though there are minor variations from year to year, all four records show peaks and valleys in sync with each other. All show rapid warming in the past few decades, and all show the last decade as the warmest.

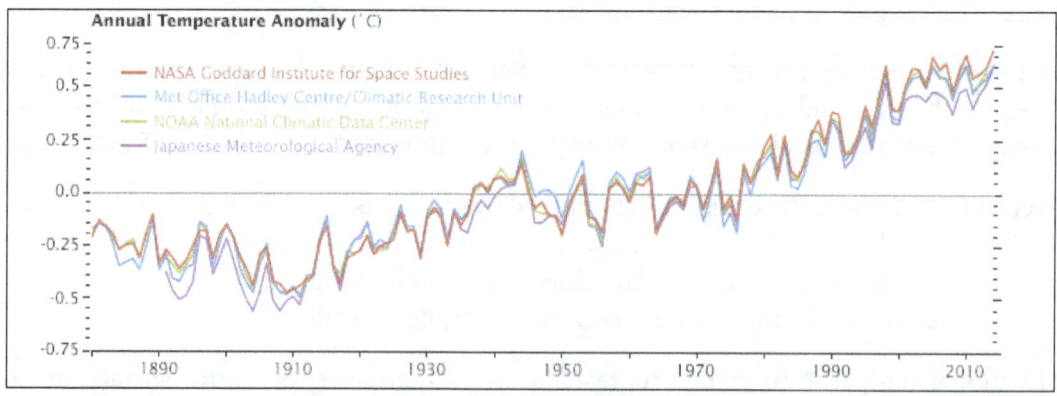

To conduct its analysis, GISS uses publicly available data from 6,300 meteorological stations around the world; ship- and buoy-based observations of sea surface temperature; and Antarctic research station measurements. These three data sets are loaded into a computer analysis program—available for public download from the GISS web site—that calculates trends in temperature anomalies relative to the average temperature for the same month during 1951-1980.

The objective, according to GISS scientists, is to provide an estimate of temperature change that could be compared with predictions of global climate change in response to atmospheric carbon dioxide, aerosols, and changes in solar activity.

As the maps show, global warming doesn't mean temperatures rose everywhere at every time by one degree. Temperatures in a given year or decade might rise 5 degrees in one region and drop 2 degrees in another. Exceptionally cold winters in one region might be followed by exceptionally warm summers. Or a cold winter in one area might be balanced by an extremely warm winter in another part of the globe.

Generally, warming is greater over land than over the oceans because water is slower to absorb and release heat (thermal inertia). Warming may also differ substantially within specific land masses and ocean basins. The graph shows the long-term temperature trends in relation to El Niño or La Niña events, which can skew temperatures warmer or colder in any one year. Orange bars represent global temperature anomalies in El Niño years, with the red line showing the longer trend. Blue bars depict La Niña years, with a blue line showing the trend. Neutral years are shown in gray, and the black line shows the overall temperature trend since 1950.

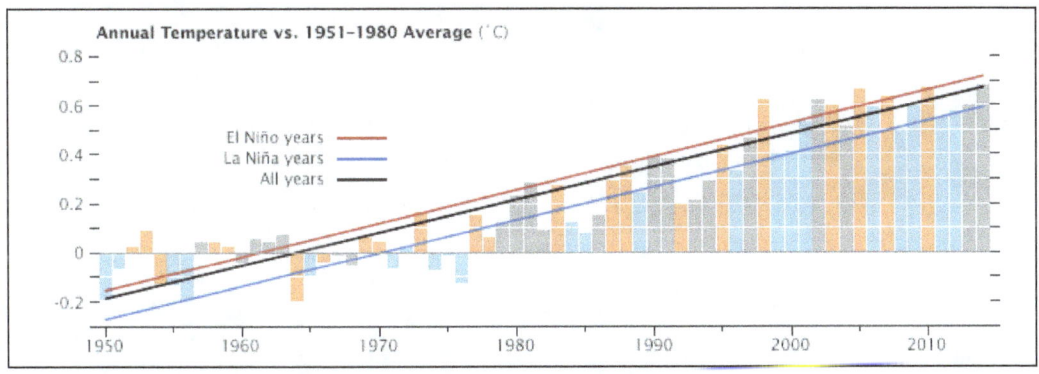

Since the year 2000, land temperature changes are 50 percent greater in the United States than ocean temperature changes; two to three times greater in Eurasia; and three to four times greater in the Arctic and the Antarctic Peninsula. Warming of the ocean surface has been largest over the Arctic Ocean, second largest over the Indian and Western Pacific Oceans, and third largest over most of the Atlantic Ocean.

Decades within the base period do not appear particularly warm or cold because they are the standard against which all decades are measured. The leveling off between the 1940s and 1970s may be explained by natural variability and possibly by cooling effects of aerosols generated by the rapid economic growth after World War II.

Fossil fuel use also increased in the post-War era (5 percent per year), boosting greenhouse gases. But aerosol cooling is more immediate, while greenhouse gases accumulate slowly and take much longer to leave the atmosphere. The strong warming trend of the past three decades likely reflects a shift from comparable aerosol and greenhouse gas effects to a predominance of greenhouse gases, as aerosols were curbed by pollution controls, according to GISS director Jim Hansen.

References

- "Greenhouse Gases | Monitoring References | National Centers for Environmental Information (NCEI)". Www.ncdc.noaa.gov. Retrieved 2019-06-06

- Toxipediagreenhouseeffectarchive, docs: healthandenvironment.org, Retrieved 11 March, 2020

- Driscoll, P.; Bercovici, D. (November 2013). "Divergent evolution of Earth and Venus: Influence of degassing, tectonics, and magnetic fields". Icarus. 226 (2): 1447–1464. Bibcode:2013Icar..226.1447D. Doi:10.1016/j.icarus.2013.07.025

- Kaltenegger, Lisa (2015), "Greenhouse Effect", in Gargaud, Muriel; Irvine, William M.; Amils, Ricardo; Cleaves, Henderson James (eds.), Encyclopedia of Astrobiology, Springer Berlin Heidelberg, p. 1018, doi:10.1007/978-3-662-44185-5_673, ISBN 9783662441848

- "G. C. SIMPSON, C.B., F.R.S., ON SOME STUDIES IN TERRESTRIAL RADIATION Vol. 2, No. 16. Published March 1928". Quarterly Journal of the Royal Meteorological Society. 55 (229): 73. 1929. Bibcode:1929QJRMS..55Q..73.. Doi:10.1002/qj.49705522908. ISSN 1477-870X

Mitigation and Management

Practices of sustainable energy, sustainable transport, greenhouse gas removal, mobile emission reduction credit, carbon dioxide removal, fossil fuel phase-out, etc. are some of the mitigation and management methods used for greenhouse gas mitigation. This chapter closely examines the greenhouse gas mitigation and management practices to provide an easy understanding of the subject.

DETECTION OF GREENHOUSE GASES USING THE PHOTOACOUSTIC SPECTROSCOPY

It is necessary to use suitable analytical techniques to identify the atmospheric components and to determinate their trace concentrations. A trace sensor of atmospheric pollutants must meet a set of fundamental requisites. High selectivity is necessary to distinguish the gas species present in a multicomponent gas mixture, such as air, and high sensitivity is essential to detect very low concentrations of substances. A large dynamic range is important to monitor the gas components at high and low concentrations using a unique instrument. In addition, a good time resolution ensures the possibility of on-line analyses controlled by a computer. Photoacoustic spectroscopy meets these requirements that enable this technique to offer important advantages in pollutant gas monitoring.

In conventional spectroscopy, the absorption of radiation is measured from the power transmitted through the sample. On the contrary, in photoacoustic spectroscopy, the absorbed radiation power is determined directly via its heat and hence the sound produced in the sample. This methodology is based on the so called photoacoustic effect which consists on the generation and detection of pressure waves (sound) inside a resonant cell, where the gas samples are placed. These samples are exposed to the incidence of modulated radiation, absorbing it at determined wavelengths. The resonant absorption of radiation generates a modulated heating in the sample and, therefore, a sound signal is produced (photoacoustic effect) and detected by highly sensitive microphones coupled, inside the cell. These microphones convert the sound signal into an electric signal, which is filtered and detected by a lock-in amplifier. Photoacoustic spectroscopy is widely used for the detection of several gases in the concentration range of ppbv and sub-ppbv. There are some types of trace gas detection systems based on continuous wave (CW) CO_2 laser, optical parametric oscillator (OPO) in combination with photoacoustic spectroscopy and quantum cascade laser (QCL). These experimental arrangements are efficient in the detection of greenhouse gases and their precursors.

Photoacoustic Spectroscopy

Currently, photoacoustic spectroscopy has been consolidated as an effective option for trace gases analysis for high sensitivity trace gases analysis (detection limits in the range of subppbv and ppbv), good selectivity, possibility of in situ measurements and continuous flow systems associated with the possibility of non-destructive analysis are attributes that allow photoacoustic to be a powerful analytical tool for gases monitoring. The photoacoustic effect consists on the generation of sound waves from the absorption of a pulsed modulated radiation. It was discovered in 1880 by Alexander Graham Bell. This discovery raised the interests of other researchers, such as John Tyndall, Wilhelm Röntgen and Lord Rayleigh. However, the lack of equipment (radiation sources, microphones, amplifiers, etc.) prevented the immediate development of this new research field, and soon the photoacoustic effect has become a mere scientific curiosity, remained virtually forgotten for over half a century.

In the late 1930s, Viengerov introduced an infrared absorption photoacoustic system to analyze gases infrared absorption. Then, Luft improved the sensitivity of this technique, allowing absorption measurements of gaseous species in the concentration range of ppmV (10^{-6}). Since the 1960s, the development of Lasers, sensitive microphones and lock-in amplifiers permitted this technique to have a great technical development. In 1968, L. B. Kerr and J. G. Atwood conducted the first photoacoustic experiments using Laser as the radiation source. Using a continuous wave CO_2 laser as excitation source, they achieved a minimum absorption coefficient of 1.2 x 10^{-7} cm^{-1} of CO_2 diluted in nitrogen. In 1971, L. B. Kreuzer, using a HeNe laser operating on 3.39 μm attained a detection limit of 10ppbv of methane diluted in nitrogen. Kreuzer, N. D. Kenyon and C. N. K. Patel used CO_2 and CO lasers as radiation sources to perform photoacoustic measurements of many trace gases. The use of a photoacoustic cell operating at in resonant mode, associated with the modulation of the excitation beam in the acoustic resonance frequencies of the cell, was introduced in 1973 by CF Dewey Jr., RD Kahn and CE Hackett. The success achieved by these pioneering studies provided a great interest in the application of photoacoustic spectroscopy in trace gases analysis in many scientific areas, especially in pollutant gas studies.

Advances in laser technology allowed the development of new infrared radiation sources continuously operating in a wide range of wavelengths, especially in the mid-infrared region. Among these new Lasers, the most promising for gas sensing are the optical parametric oscillator (OPO) and the quantum cascade laser (QCL). The combination of these new radiation sources with photoacoustic cell optimized settings (differential photoacoustic cell, intra-cavity arrangements or multi-pass) enabled major advances in trace gases photoacoustic detection.

Photoacoustic effect in Gases

The photoacoustic signal generated and detection in gases been studied mainly by Kreuzer and revised and expanded by several authors. Molecular absorption of photons results in the excitation of molecular energy levels (rotational, vibrational and electronic) degrees freedom. The excited state loses its energy by radiative processes, such as spontaneous or stimulated emission, and by collisional relaxation, in which the state energy is transformed into translational one.

In the case of vibrational excitation, radiative emission and chemical reactions do not play an important role, because the radiative lifetimes of vibrational levels are long compared with the time needed for collisional deactivation at pressures used in photoacoustic. Furthermore, the photon energy is too small to induce chemical reactions. For 1 atm pressure, the vibrational-translational non-radiative decay time is typically around 10^{-6}-10^{-9}s, whereas the radiative lifetime is between 10^{-1} and 10^{-3}s. Thus, in practice, the absorbed energy is completely released as heat, appearing as translational (kinetic) energy of the gas molecules.

By modulating the intensity or the wavelength of the incident radiation, the sample local heating and expansion become periodic. If radiation intensity is modulated (without optical saturation), the heat density in the sample (H) is directly proportional to the volumetric density of molecules (N), to the absorption cross section of the absorbing molecule (σ) and to the incident laser radiation intensity (I_0). Therefore, the gas heat production is given by:

$$H(\mathbf{r},t) = N\sigma I_0 e^{i\omega t}$$

The sample pressure oscillations (p) are related to the heat production by the following equation:

$$\nabla^2 p - \frac{1}{v^2}\frac{\partial^2 p}{\partial t^2} = -\frac{\gamma-1}{v^2}\frac{\partial H}{\partial t}$$

Where v, γ, and H are the sound velocity, the adiabatic co-efficient of the gas, and the heat density deposited in the gas by light absorption, respectively. All dissipative terms of heat diffusion to the cell walls and viscosity were despised.

For many applications, this simplified approach is sufficient, although in some cases, more stringent treatment that takes into account relaxation effects is necessary. Since the amplitude of the acoustic waves produced depends on both the nature of the absorbing gases and their concentrations, the photoacoustic detection allows qualitative and quantitative analysis of gas mixtures.

The acoustic wave is generated in places where light is absorbed by the monitored species. The photoacoustic signal $S(\lambda)$ produced by a single absorbing gaseous species diluted in a non absorbing gas can be expressed as:

$$S(\lambda) = CP(\lambda)N_{tot}c\sigma(\lambda) = CP(\lambda)\alpha(\lambda)c$$

Where C is the cell constant, which depends on the cell geometry, the microphone responsivity and on the nature of the acoustic mode, $P(\lambda)$ is the laser power at the wavelength λ, N_{tot} is the total density of molecules, considering a pressure of 1013hPa and a temperature of 20 °C, $N_{tot} \approx 2.5 \times 10^{19}$ molecules/cm^3, c and $\sigma(\lambda)$ are, respectively, the concentration (mole fraction) and the absorption cross section of the absorbing molecule and $\alpha(\lambda) = N_{tot} . c\sigma(\lambda)$ is the so called gas absorption coefficient (cm^{-1}).

Using equation, it is possible to determine the minimum detectable gas concentration in the photoacoustic spectrometer, through:

$$c_{min} = \frac{S_{min}}{N_{tot}CP\sigma} = \frac{S_{min}}{CP\alpha}$$

where $S_{min}(\lambda)$ is the minimum detectable signal, which is usually measured by passing a flow of non-absorbing inert gas (usually nitrogen or synthetic air) in the photoacoustic cell. The minimum signal is produced by the various sources of noise, which are always present in photoacoustic detection, determining its limitations. The most common sources of noise in photoacoustic systems include acoustic signals caused by the windows heating, the absorption and scattering of radiation on the cell resonator walls, by molecules adsorbed on them, the noise caused by the gas flow and electronic noise.

We can notice that the signal is obtained at a given wavelength, which is specific for the molecule we wish to detect. Moreover, the obtained signal is directly proportional to the concentration of the absorbing gas. Thus, it is possible to obtain the concentration as a function of the generated signal.

In order to determine the concentration of various gaseous species in a multicomponent sample, the photoacoustic signal can be obtained for different wavelengths, corresponding to the absorption of each analyzed component. In this case, the signal can be obtained by the relation:

$$S(\lambda_i) = CP(\lambda_i)N\sum_{j=1}^{n}c_j\sigma_{ij}$$

with i = 1, 2, ..., m and j = 1, 2, ..., n e m ≥ n. Here, $P_i = P(\lambda_i)$ represents the laser power at wavelength λ_i and c_j is the concentration of the component $_j$ with absorption cross section σ_{ij} at λ_i. This equation solution for each gas species concentration is given by:

$$c_j = \frac{1}{CN}\sum_{i=1}^{m}(\sigma_{ij})^{-1}\left(\frac{S_i}{P_i}\right)$$

Where $(\sigma_{ij})^{-1}$ is the inverse matrix of the matrix σ_{ij}. The success of this method in multicomponent mixture analysis strongly depends on the nature of the matrix σ_{ij}. The trivial case is represented by a diagonal matrix (σ_{ij}), where a set of wavelengths λ_i can be selected and the absorption by a single gas component occurs in each λ_i. However, this ideal case without interference between the absorptions due to different gas species in a multicomponent mixture hardly occurs. Instead, considerable effort is needed to discriminate the various components, as well as to identify any unexpected component. Iterative algorithms using least squares regression and based on prior knowledge of the absorption cross sections (σ_{ij}) have been developed to fit the photoacoustic spectrum of multicomponent samples.

Lasers as Radiation Sources

The selectivity required for the identification of the components of a gaseous mixture is achieved in spectroscopy when lasers are employed. Due to its high radiance, narrow linewidth and wide spectral range of emission, lasers are indeed suitable for trace gas monitoring. They allow a selective detection of different gases in trace level concentration (lower limit: parts per trillion by volume) even when there is overlapping of spectrum lines from different molecules. It is known that for molecules, absorption spectra in the near infrared (NIR) and the infrared (IR) ranges work as fingerprints, that is, they are unique for each molecular species. Currently, there are commercially available lasers that emit in these regions, for example, semiconductor lasers (DFB diode lasers) and gas lasers (CO and CO_2), respectively. The great advantage of using these light sources is the resulting high spectral resolution.

Promising new sources of light are the quantum cascade lasers (QCL). These lasers emit in the mid-IR range, where the molecules have higher absorption coefficients, and operate at room temperature. Since the photoacoustic signal intensity is also a function of the molecular absorption coefficients, the low power these lasers emit is hence compensated. Another important feature and advantage of these lasers is the high spectral resolution. Furthermore, the emission wavelength can be selected to match two, or more, absorption lines of a given molecule. Third generations of extremely promising lasers are the Optical Parametric Oscillator lasers (OPO).

CO_2 Laser

The use of tunable CO_2 lasers, through the scanning of its emission wavelengths (9-11$\dashv$ m) ensures the exploration of the so-called molecular fingerprint, enabling the identification and simultaneous monitoring of several gaseous compounds with a single instrument. Figure shows the absorption spectra of ozone (O_3), ammonia (NH_3) and ethylene (C_2H_4), for the emission range of a CO_2 laser. As gas lasers have high power (> 10 watts) and the signal intensity is directly proportional to the emitted light power, its use results in sensitivities in the range of pptv. Nevertheless, these lasers have two disadvantages, they are large and expensive and thus its use is limited to laboratories.

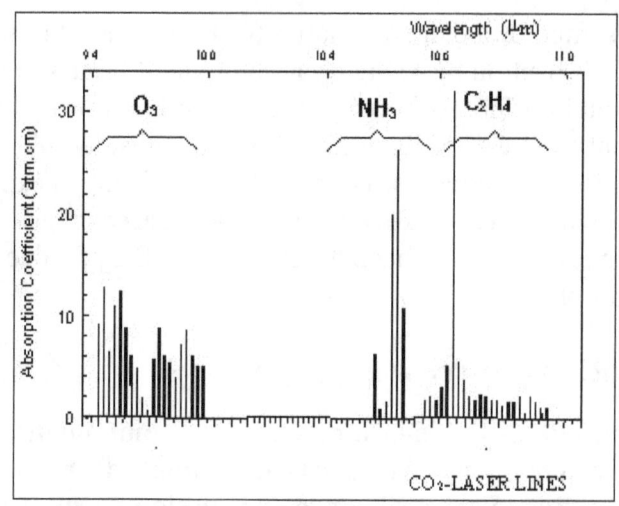

Fingerprints of ozone (O_3), ammonia (NH_3) and ethylene (C_2H_4) corresponding to the emission lines of a CO_2 laser.

Quantum Cascade Laser (QCL)

The quantum cascade laser (QCL), introduced by Faist, represents an excellent source of radiation for trace gas monitoring. Although providing output power well below the CO_2 laser (a few mW), this source has some advantages such as compact size, continuous emission, high spectral resolution and the possibility of being operated near room temperatures.

In addition, they can be manufactured to operate in a wide range of wavelengths, from 3 to 24 μm. Among its applications, we highlight the environmental monitoring, industrial process control, besides chemical and biomedical applications.

The QCL laser coupled with a photoacoustic detector has been successfully used to measure the concentration of different gases which absorb radiation in the mid- infrared spectral region, such as ozone, ammonia, and NO_2 and N_2O gases.

This type of laser is part of the family of semiconductor lasers, with the particularity of using quantum transitions within the same band. A quantum cascade laser comprises a series of alternate thin layers of two different materials. This configuration enables different electrical potentials to be established across the device, so that electrons can get trapped in these sites, called quantum wells. Thus, a series of sub-bands with different energies is created inside the conduction band.

When stimulated, the electrons undergo transitions and quantum tunneling to a lower energy sub-band, and consequently photons are emitted. A single electron can perform several transitions, that is, issue multiple photons.

Another important characteristic of quantum cascade lasers is that the wavelength can be determined by the thickness of the layers, rather than being determined by the energy difference between bands. The thickness and refractive index change by setting different temperatures turning the wavelength.

The Optical Parametric Oscillator Laser (OPO)

An interesting optical process that has been used to produce near- and mid-infrared radiation is the optical parametric generation. In 1962, Armstrong et al and Kroll described the fundamental theory of the optical parametric generator and three years later, Giordmaine and Miller demonstrated the operation of an optic parametric oscillator. When an optically nonlinear crystal is submitted to electromagnetic fields of high density of energy such as those present in pulsed lasers, the electrons respond with significant displacement that gives rise to the contribution of the second-order nonlinear component for the electric polarization of a nonlinear medium. The mixing of two electromagnetic waves under condition of nonlinear polarization produces parametric effects such as second harmonic generation (frequency doubling), sum frequency generation and difference frequency generation. In the latter case, when the nonlinear crystal is pumped by two input photons with wavelength at λ_p (pump photon) and λ_s (signal photon, $\lambda_s > \lambda_p$), the signal photon stimulates the conversion of a pump photo into a new signal and an idler photon with wavelength at $\lambda_i^{-1} = \lambda_p^{-1} - \lambda_s^{-1}$ (fulfilling the energy conservation principle). Of course, the efficiency of such conversion is limited to phase match between the pump and signal waves.

This process of increasing the number of signal photons is known as optical parametric amplification (OPA) and thus it fundamentally differs from the amplification mechanism in laser since no population inversion and excited states take place. Theoretically, the remarkable advantage of OPA is the infinity possibility of combining two waves that generates a third wave with different wavelength. This fulfills the expected desirable feature of an excitation source for analytical spectroscopy application that is the broad wavelength tunability. Therefore, this makes the optical parametric phenomena of great significance to spectroscopy application in the sense of allowing selective trace gas detection and thus analysis of multicomponents samples, such as the air. Although the finite width of the emission line may reduce the selectivity, for atmosphere application the broadening of the line width of detected specie due to the atmospheric pressure for itself reduces the significance of an ultra-narrow line width of an applied source.

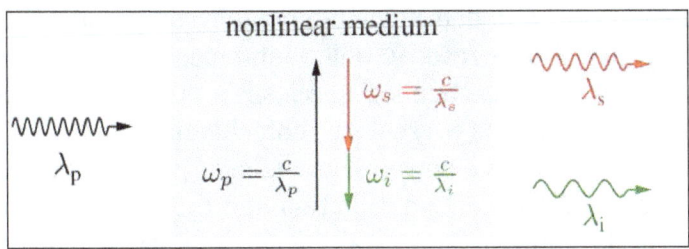

Optical parametric process: The incident pump radiation with wavelength λ_p and circular frequency at ωp is converted into signal and idler radiations with wavelength

λ_s and λ_i and circular frequency at ω_s and ω_i, respectively. c is the velocity of the light in the medium.

The efficiency of the nonlinear conversion depends on the phase matching of the pump and signal waves. The dispersion of electromagnetic wave propagating through a crystal is directly related to the refractive index of the media that is different for each wavelength. Therefore the pump and signal wave move in and out of phase relatively to each other, limiting the quantity of generated signal photons. Consequently, initially the first issue was to find materials that provide phase matching for at least two wavelengths. The use of birefrigent crystalline materials was the first key to fix the relative phase of the pump and signal waves. However, the available range of wavelengths that satisfies the phase-matching condition is limited to the variability of birefrigent crystalline materials. More recently, the use of periodically poled lithium niobate (PPLN) has overcome this restriction. PPLN chips display an engineered inverted orientation of lithium niobate crystals that promotes a quasi-phase-matched combination of the pump and signal waves compensating the phase mismatch present in parametric interaction.

For the optic parametric oscillator (OPO), the crystal is initially pumped by photons of only one wavelength λ_p. Based on the fundamental quantum uncertainty in the electric field (quantum noise), a pump photon (λ_p) propagating in a nonlinear optical crystal spontaneously breaks up spontaneously into two lower-energy photons with wavelength at λ_s (signal photon) and λ_i (idler photon). This optic parametric process is called optic parametric generation (OPG). Afterwards, the created signal wave mixes with the pump wave under condition of nonlinearity resulting in new signal and idler waves (stimulated generation). To increase the number of signal and idler photons, the crystal is placed within an optic cavity formed by two mirrors (optical resonator). Single resonance is achieved when the signal wave is reflected back and forth in the optic cavity.

Detection of Greenhouse Gases

CO_2 Laser Experimental Setup

The methodologies based on photothermal techniques, mainly CO_2 laser photoacoustic spectroscopy, have suitable characteristics to detect trace gas, as high sensitivity and selectivity and possibility of in situ measurements. A CO_2 laser based photoacoustic spectrometer can be used to detect volatile organic compounds (VOCs) emissions, such as ethylene. Ethylene is a reactive pollutant, since it is an unsaturated organic compound. For this reason, this chemical species is a precursor for the generation of the tropospheric ozone, which is present in photochemical smog and directly affects human health. Besides, ozone is a powerful greenhouse gas, whose formation is greatly potentiated by the incidence of sun radiation and the presence of nitrogen oxides (NO_x). According to the Intergovernmental Panel of Climatic Changes (IPCC), ozone has a positive radiative forcing of about 0.35 W/m², being, therefore, an important source of global warming.

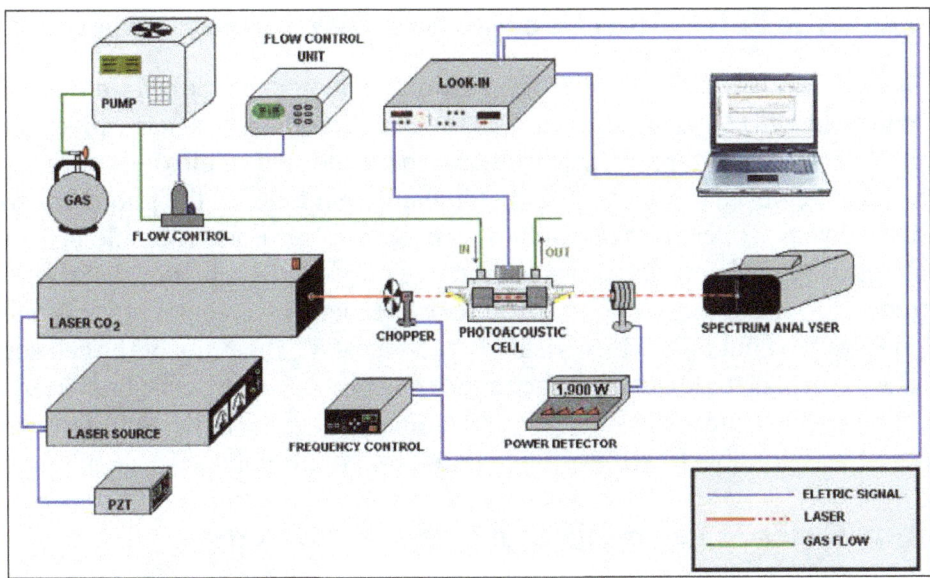

Scheme of the photoacoustic experimental setup.

To guarantee a refined detection of ethylene, the photoacoustic spectrometer is daily calibrated by submitting the cell to a flow of a certified mixture. This measurement was carried out using a certified gas mixture of 1.1 ppmV ethylene in N_2 flowing into the cell at a rate of 83.3 sccm (standard cubic centimeter). The acoustic signal is detected by a microphone that generates an electric signal. This electric signal is pre-amplified and then detected by a lock-in amplifier (Stanford SR850) with a time constant of 300 ms. The lock-in response is registered in a microcomputer. A continuous wave CO_2 infrared laser (LTG, model LTG150 626G), tuneable over about 80 different lines between 9.2 and 10.6 μm, with a power of 1.9W at the emission line 10P(14) (10.53 ⊣m), by internal PZT (Piezoelectric Transducer), is employed as the excitation source. At this power level, no saturation effects of the photoacoustic signal were observed. These lines can be swept by a step motor controlled by a microcomputer. Within this spectral region, many small molecules show a unique fingerprint. The photoacoustic instrument used has been developed for the detection of small concentrations of gases. All the measurements and the sample collection are made at room temperature. Therefore, the analysis of these samples is made for a number of n different species, rather than just one. This was accomplished by measuring the photoacoustic (PA) signal $S(\lambda_i)$ at a set of wavelengths λ_i (i = 1, 2... m) chosen on the basis of the absorption spectra of the individual components to be detected. These individual absorption spectra were obtained from the HITRAN-PC database, which calculates the absorption cross sections (σ) of a given molecule at different wave numbers $k_i = 1/\lambda_i$ in a given interval. Thus, the expression used to determine the concentrations of a given component in the multicomponent gas mixture is the equation. The absorption cross section σ_{ij} is related to the photoacoustic generation efficiency of each gas component for each CO_2 laser line. The sum is taken over the n components present in the sample.

Photoacoustic Cell Calibration and Sensitivity Measurements

The calibration and sensitivity measurements of the photoacoustic cell were performed by obtaining the cell coupling constant C in the equation. This was performed by taking a 1.1 ppmV certified mixture of ethylene in N_2 and diluting it in nitrogen until the least concentration achieved (about 16 ppbv). A linear dependence of the photoacoustic signal on the ethylene concentration could be proven and this linearity could be extended to ppmV levels. The absorption cross section σ of ethylene is well known at the 10P(14) (949.51cm^{-1}) CO_2 laser line ($\sigma = 170 \times 10^{-20}$ cm^2). Hence, the C constant value was then obtained from the equation, which yielded 40.2 V.cm/W. The unity of the cell coupling constant was furnished by the manufacturer of our photoacoustic cell (Prof. Markus W. Sigrist). Recent measurements made in greenhouse gas SF_6, indicated that using the CO_2 laser, it was possible to achieve a detection limit of 20 ppbv.

Quantum Cascade Laser (QCL) Experimental Setup

With the recent development of quantum-cascade lasers (QCLs), compact, low-cost, solidstate radiation sources are available, covering the important infrared (IR) region with specific molecular absorption lines. In addition, spectral regions can be selected in which water vapor has a very low absorption coefficient, known as atmospheric windows. Another important advantage of QCLs in practical applications is that they work near room temperature, whereas diode lasers such as lead salt lasers, which emit in the fundamental IR region, have to be cryogenically cooled. Recent applications of QCLs clearly indicate their potential as tunable light sources in the mid-infrared, especially between 3 - 13 μm, with strong fundamental absorption bands. Current interest is based on the lack of other convenient coherent laser sources. In fact, it can be expected that QCLs will open new possibilities for real-time diagnostics of various molecular species in the 3-5 μm and 8-13 μm atmospheric windows. Pulsed quantum-cascade distributed-feedback (QC-DFB) lasers provide quasi room temperature operation, combined with a high spectral selectivity and sensitivity, real time measurement capabilities, robustness, and compactness.

For this reason, QCLs are ideal for the development of compact trace gas analyzers that are also suitable for field measurements. In recent years the detection of a series of important trace gases has been demonstrated with these devices. By way of illustration we report on measurements of sulphur hexafluoride and methane with a homemade Laser Photoacoustic Spectrometer equipped with QC lasers and a Differential Photoacoustic Cell. The motivation of our research comes from the need for simple, sensitive, and spectrally selective devices for measuring traces of greenhouse gases in agriculture, automobile exhaust monitoring, power distribution facilities, cattle breeding and chemistry industries. The experimental set up employed in the detection limit determination of the analyzed gases is illustrated in figure.

As radiation source, a pulsed quantum cascade is normally used. In this experiment two quantum cascade lasers were employed separately, each laser emission band matching the absorption lines of one of the specified molecules. The laser used in the detection of CH_4, emits in the range of 7.71 - 7.88 µm and can reach a power of 5.6 mW (at lowest operating temperature of the laser), the one employed in the detection of SF_6, emits in the range of 10.51 - 10.56 µm and can reach a power of 3.7 mW.

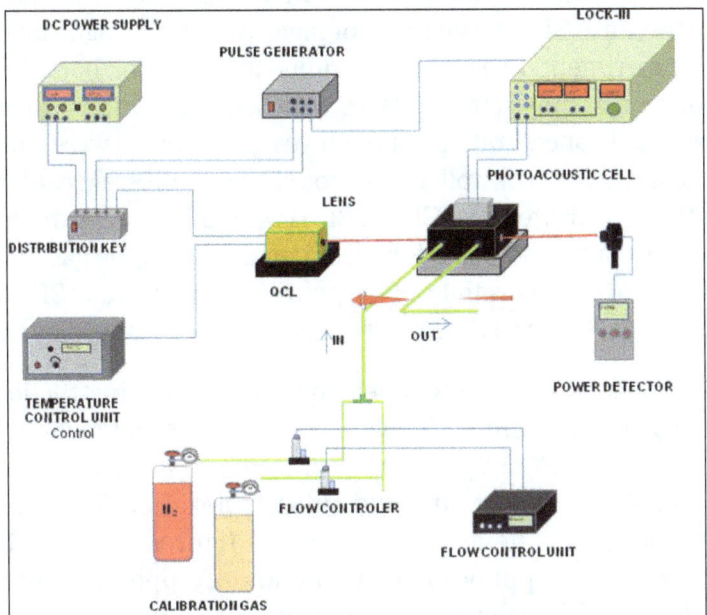

Experimental set up using quantum cascade laser.

The laser emission lines are given according to the diode temperature, which is determined by a temperature control unit. The pulsed QCL light beam, with a repetition rate of 400kHz and a pulse duration of 50ns (duty cycle of 2%), is gated by an external transistor-totransistor logic (TTL) signal at 3.8kHz to excite the first longitudinal acoustic mode of the resonant differential photoacoustic cell. A germanium lens (focus ~30.7 mm and diameter ~10.35mm) is employed to focus the QCL radiation through the cell. The cell has two resonant cylindrical tubes (5.5 mm in diameter and 4 cm in length) on whose edges are arranged acoustic buffers which reduce noise caused by gas turbulence and background signal produced by the heating in the cell windows when these are exposed to the radiation. The gas flow streams through both pipes and noise and background are equally detected by the microphones placed on each of them, but only the microphones placed in the tube crossed by the laser beam detect the pressure change induced by the absorption of modulated radiation in the gaseous sample containing the molecules under consideration. Thus the photoacoustic signal is obtained by simple differentiation of the signal produced by the microphones in the two tubes.

The laser power is monitored by a power detector (OPHIR, 3A-SH-ROHS) and the gases flows are controlled by electronic mass flow controllers (model MKS, 247), one of de 50 sccm and one of 300sccm. The PA data analysis was performed by the lock-in technique using a lock-in amplifier (model Stanford SR_850 DSP) with a set data acquisition time constant of 300ms.

Calibration and Sensitivity Measurements

The following concentration measurements were performed keeping the temperature of each laser constant. At these emission lines the lasers power was 0.8 mW, feed current of 26.2 mA, for the laser used to measure CH_4, and 1.12 mW, feed current of 25.3 mA, for the laser used to measure SF_6. In such type of measurement, a high stability is observed during the entire experiment. In order to determine the detection boundaries of the gases of interest, a dilution of standard mixtures were carried out. Dilution experiments are depicted in figures. Small concentrations of the investigated gases were synthesized by using two electronic mass-flow controllers, one for N_2 (with full scale control of 200 sccm) and another for the investigated gas (CH_4 or SF_6) (with full scale control of 50 sccm). The electronic mass-flow controllers were connected in parallel to the gas inlet of the photoacoustic cell. The initial concentrations of 4.5 ppmv CH_4 and 5 ppmv SF_6 was diluted with pure nitrogen (zero gas) down to the lowest concentrations detected by the system.

The acoustic and electronic noise was determined by blocking the laser light while keeping all other devices running. The value of the noise signal was typically 0.300 μV. As expected $S(\lambda) = CP(\lambda)N_{tot}c\sigma(\lambda) = CP(\lambda)\alpha(\lambda)c$, a linear dependence of the photoacoustic signal on the methane and sulphur hexafluoride concentration was found. The fitted straight line are also shown in figures. The smallest measured concentrations were of 1.5 ppmv for methane and 49 ppbv for sulphur hexafluoride. Although the smallest concentration of methane detected was of 1.5 ppmv, the strong linear slope of the fitted straight line allows us to estimate that the instrumentation has the sensitivity to detect concentration changes smaller than 1.5 ppmv. In recent measurements made with methane, it was possible to achieve an experimental detection limit of 50 ppbv, for this gas. It is possible to estimate a detection limit of 30 ppbv methane, by extending the straight line until the noise limit, at a signal to noise ratio, in single pass.

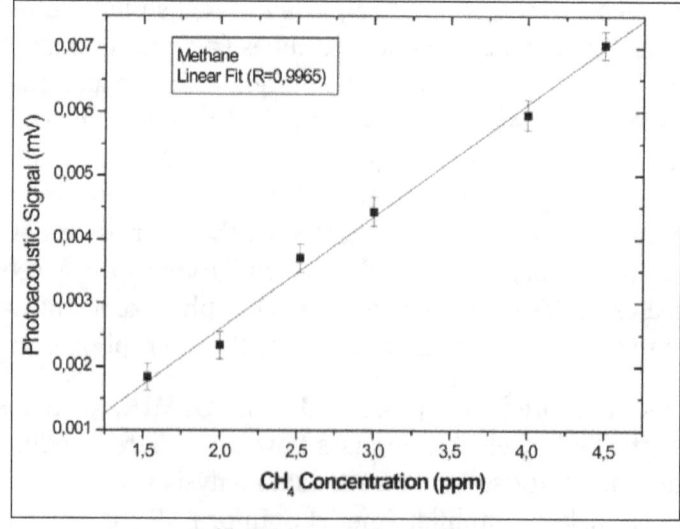

Calibration curve for methane.

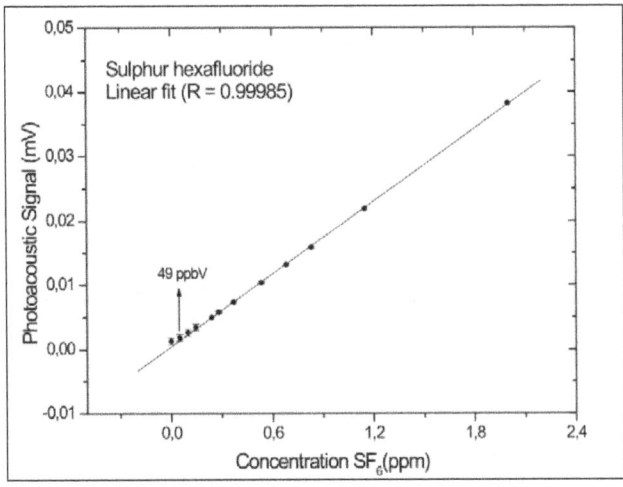

Calibration curve for sulphur hexafluoride.

The photoacoustic spectroscopy with quantum cascade lasers has proved to be extremely efficient for the detection of greenhouse gases, being sensible and selective. In the case of the greenhouse gas methane, this methodology allows measurements in anthropogenic sources that emits methane in concentration higher than 1.5 ppmv and also atmospheric measurements, once it is estimated that the current average concentration of this gas in the atmosphere is of 1.7 ppmv. For sulphur hexafluoride the method is suitable for concentrations greater than 49 ppbv, a concentration already detected by conventional equipments. Another important greenhouse gas is the nitrous oxide (N_2O) which can also be detected by quantum cascade laser (QCL), whose detection limit of 14 ppbv was obtained for this gas.

OPO Laser Experimental Setup

Figure shows a schematic diagram of an experimental setup for an OPO in a simple grazing-incidence grating configuration (GIOPO). Two gold mirrors (M1 and M2) are used to produce an optical resonator. The pump light (1064 nm) is put into the resonator by 45° incidence on a third mirror (M3) coated for high reflection at 1064 nm and high transmission for signal and idler waves. A coated highly transmitting at 1064 nm lens (L1) is used to focus the pump beam at the middle of the PPLN crystal (C1) length. In the GIOPO configuration a grating (600/mm groove density) (G1) is placed at grazing incidence relative to the cavity axis. The grating serves as dispersing element. The diffracted first order off the grazing is reflected back into the cavity by M2 and used as injection seed. The zero order is out coupled from the resonator to be used as exciting light for photoacoustic spectroscopy. Before the beam reaches the photoacoustic detector, a highly reflecting mirror at 1064 nm is employed to eliminate the pump wave and a germanium element (Ge) is used to filtering the signal wave from the beam. This OPO configuration results in a typical linewidth of about 0.1 cm^{-1} and the idler wave covers a wavelength range between 2.4 and 4 μm with power average of some hundreds of miliwatts by either tuning the mirror M2 or changing the temperature of the crystal.

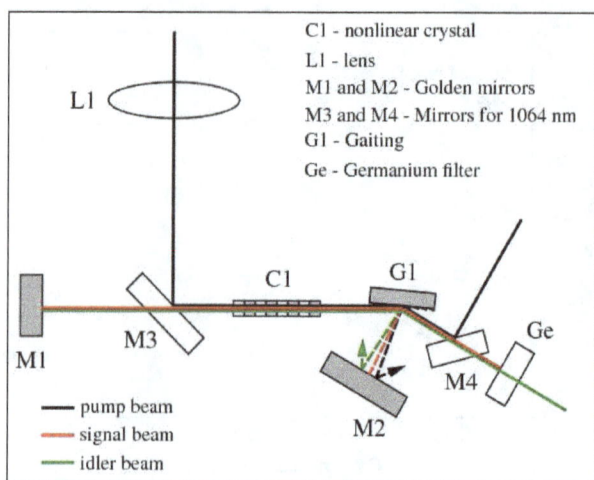

Schematic setup for OPO with in a grazing-incidence grating configuration.

Owing to the wavelength tunability of the OPO, in recent years it has been shown the feasibility of using optical parametric oscillator as light source for trace gas detection of several chemical species of environmental appeal. Limits of part per billion for greenhouse gases have been demonstrated when OPO is combined with photoacoustic detection methods. Concentration limit of 60 ppbv (part per billion by volume) and 20 ppbv were estimated for nitrous oxides (N_2O) and methane (C_2H_4), respectively, when the OPO radiation is in amplitude modulated. The sensitivity of the detection for methane can be improved when the technique of wavelength modulation of the OPO radiation, combined with multipass configuration, is employed. This modulation mode has the advantage of the suppression of the baseline caused by hits of the cell windows and the reflection on the mirrors. According to Nd et all, an ultimate sensitivity of 136 parts in 10^{12} for methane was estimated when the technique of wavelength modulation is used.

Since the increase rate of greenhouse in the atmospheric air is higher than the sensitivity of an OPO setup, it gives rise in this way the possibility of using OPO devices to monitor the annual change of pollutants in atmospheric air. Detection of N_2O in ambient air was already carried out using photoacoustic spectroscopy. Applying the photoacoustic spectroscopy in combination with a pulsed grazing-incident optical parametric oscillator, concentrations of 311 ppbv were found for ambient samples collected at nearby roads.

GREENHOUSE GAS REMOVAL

Greenhouse gas removal projects are a type of climate engineering that seek to remove greenhouse gases from the atmosphere, and thus they tackle the root cause of global warming. These techniques either directly remove greenhouse gases, or alternatively seek to influence natural processes to remove greenhouse gases indirectly.

Carbon Sequestration

A wide range of techniques for carbon sequestration exist. These range from ideas to remove CO_2 from the atmosphere (carbon dioxide air capture), flue gases (carbon capture and storage) and by preventing carbon in biomass from re-entering the atmosphere, such as with Bio-energy with carbon capture and storage (BECCS).

Pyrogenic Carbon Capture and Storage

Pyrogenic carbon capture and storage is a promising technology of greenhouse gas removal. Pyrolysis is described by Werner et al. as "the thermal treatment of biomass at 350 °C–900 °C in an oxygen-deficient atmosphere. Three main carbonaceous products are generated during this process, which can be stored subsequently in different ways to produce [negative emissions]: a solid biochar as soil amendment, a pyrolytic liquid (bio-oil) pumped into depleted fossil oil repositories, and permanent-pyrogas (dominated by the combustible gases CO, H_2 and CH_4) that may be transferred as CO_2 to geological storages after combustion."

A different procedure has been proposed by Esrafilzadeh et al. that makes it possible to create bio char under room temperature, using liquid metal electrocatalysts.

Chlorofluorocarbon Photochemistry

Atmospheric chlorofluorocarbon (CFC) removal is an idea which suggests using lasers to break up CFCs, an important family of greenhouse gases, in the atmosphere.

Methane Removal

Methane potentially poses major challenges for remediation. It is around 20 times as powerful a greenhouse gas as CO_2. Large quantities may be outgassed from permafrost and clathrates as a result of global warming, notably in the Arctic.

There are existing climate engineering proposals. Methane is removed by several natural processes, which can be enhanced.

- Chemical decomposition: Reaction with hydroxyl radicals produced from photochemical decomposition of ozone in the stratosphere.

- Biological decomposition: By methanotrophs in soils and water.

SUSTAINABLE ENERGY

Sustainable energy is the practice of using energy in a way that "meets the needs of the present without compromising the ability of future generations to meet their own needs."

Meeting the world's needs for energy in a sustainable way is widely considered to be one of the greatest challenges facing humanity in the 21st century. Worldwide, nearly a billion people lack access to electricity, and around 3 billion people rely on smoky fuels such as wood, charcoal or animal dung in order to cook. These and fossil fuels are a major contributor to air pollution, which causes an estimated 7 million deaths per year. Production and consumption of energy emits over 70% of human-caused greenhouse gas emissions.

Proposed pathways for limiting global warming to 1.5 °C describe rapid implementation of low-emission methods of producing electricity and a shift towards more use of electricity in sectors such as transport. The pathways also include measures to reduce energy consumption; and use of carbon-neutral fuels, such as hydrogen produced by renewable electricity or with carbon capture and storage. Achieving these goals will require government policies including carbon pricing, energy-specific policies, and phase-out of fossil fuel subsidies.

When referring to methods of producing energy, the term "sustainable energy" is often used interchangeably with the term "renewable energy". In general, renewable energy sources such as solar, wind, and hydroelectric energy are widely considered to be sustainable. However, particular renewable energy projects, such as the clearing of forests for production of biofuels, can lead to similar or even worse environmental damage when compared to using fossil fuel energy. There is considerable controversy over whether nuclear energy can be considered sustainable.

Moderate amounts of wind and solar energy, which are intermittent energy sources, can be integrated into the electrical grid without additional infrastructure such as grid energy storage. These sources generated 7.5% of worldwide electricity in 2018, a share that has grown rapidly. As of 2019, costs of wind, solar, and batteries are projected to continue falling.

Buildings in the Solar Settlement at Schlierberg incorporate rooftop solar panels and are built for maximum energy efficiency. As a result, they produce more energy than they consume.

The concept of sustainable development was described by the World Commission on Environment and Development in its 1987 book Our Common Future. Its definition of "sustainability", now used widely, was, "Sustainable development should meet the needs of the present without compromising the ability of future generations to meet their own needs."

In its book, the Commission described four key elements of sustainability with respect to energy: the ability to increase the supply of energy to meet growing human needs, energy efficiency and conservation, public health and safety, and "protection of the biosphere and prevention of more localized forms of pollution." Various definitions of sustainable energy have been offered since then which are also based on the three pillars of sustainable development, namely environment, economy, and society.

- Environmental criteria include greenhouse gas emissions, impact on biodiversity, and the production of hazardous waste and toxic emissions.

- Economic criteria include the cost of energy, whether energy is delivered to users with high reliability, and effects on jobs associated with energy production.

- Socio-cultural criteria include the prevention of wars over the energy supply (energy security) and long-term availability of energy.

The organizing principle for sustainability is sustainable development, which includes the four interconnected domains: ecology, economics, politics and culture.

Current Status

Providing sustainable energy is widely viewed as one of the greatest challenges facing humanity in the 21st century, both in terms of meeting the needs of the present and in terms of effects on future generations. Bill Gates said in 2011:

> "If you gave me the choice between picking the next 10 presidents or ensuring that energy is environmentally friendly and a quarter as costly, I'd pick the energy thing."

Worldwide, nearly a billion people do not have access to electricity, and around 3 billion people rely on dirty fuels for cooking. Air pollution, caused largely by the burning of fuel, kills an estimated 7 million people each year. The United Nations Sustainable Development Goals call for "access to affordable, reliable, sustainable and modern energy for all" by 2030.

Energy production and consumption are major contributors to climate change, being responsible for 72% of annual human-caused greenhouse gas emissions as of 2014. Generation of electricity and heat contributes 31% of human-caused greenhouse gas emissions, use of energy in transportation contributes 15%, and use of energy in manufacturing and construction contributes 12%. An additional 5% are released through processes associated with fossil fuel production, and 8% through various other forms of fuel combustion. As of 2015, 80% of the world's primary energy is produced from fossil fuels.

In developing countries, an estimated 2.5 billion people rely on traditional cookstoves and open fires to burn biomass or coal for heating and cooking. This practice causes harmful local air pollution and increases danger from fires, resulting in an estimated 4.3 million deaths annually. Additionally, serious local environmental damage, including desertification, can be caused by excessive harvesting of wood and other combustible material. Promoting usage of cleaner fuels and more efficient technologies for cooking is therefore one of the top priorities of the United Nations Sustainable Energy for All initiative. Thus far, efforts to design clean cookstoves that are inexpensive, powered by sustainable energy sources, and acceptable to users have been mostly disappointing.

Proposed Pathways for Climate Change Mitigation

Cost–benefit analysis work has been done by a disparate array of specialists and agencies to determine the best path to decarbonizing the energy supply of the world. The IPCC's 2018 Special Report on Global Warming of 1.5 °C says that for limiting warming to 1.5 °C and avoiding the worst effects of climate change, "global net human-caused emissions of CO_2 would need to fall by about 45% from 2010 levels by 2030, reaching net zero around 2050." As part of this report, the IPCC's working group on climate change mitigation reviewed a variety of previously-published papers that describe pathways (i.e. scenarios and portfolios of mitigation options) to stabilize the climate system through changes in energy, land use, agriculture, and other areas.

Bangui Wind Farm in the Philippines.

Workers construct a solar panel array structure in Malawi.

The pathways that are consistent with limiting warning to approximately 1.5 °C describe a rapid transition towards producing electricity through lower-emission methods, and increasing use of electricity instead of other fuels in sectors such as transportation. These pathways have the following characteristics (unless otherwise stated, the following values are the median across all pathways):

- Renewable energy: The proportion of primary energy supplied by renewables increases from 15% in 2020 to 60% in 2050. The proportion of primary energy supplied by biomass increases from 10% to 27%, with effective controls on whether land use is changed in the growing of biomass. The proportion from wind and solar increases from 1.8% to 21%.

- Nuclear energy: The proportion of primary energy supplied by nuclear power increases from 2.1% in 2020 to 4% in 2050. Most pathways describe an increase in use of nuclear power, but some describe a decrease. The reason for the wide range of possibilities is that deployment of nuclear energy "can be constrained by societal preferences."

- Coal and oil: Between 2020 and 2050, the proportion of primary energy from coal declines from 26% to 5%, and the proportion from oil declines from 35% to 13%.

- Natural gas: In most pathways, the proportion of primary energy supplied by natural gas decreases, but in some pathways it increases. Using the median values across all pathways, the proportion of primary energy from natural gas declines from 23% in 2020 to 13% in 2050.

- Carbon capture and storage: Pathways describe more use of carbon capture and storage for bioenergy and fossil fuel energy.

- Electrification: In 2020, around 20% of final energy use is provided by electricity. By 2050, this proportion more than doubles in most pathways.

- Energy conservation: Pathways describe methods to increase energy efficiency and reduce energy demand in all sectors (industry, buildings, and transport). With these measures, pathways show energy usage to remain around the same between 2010 and 2030, and increase slightly by 2050.

Renewable Energy Sources

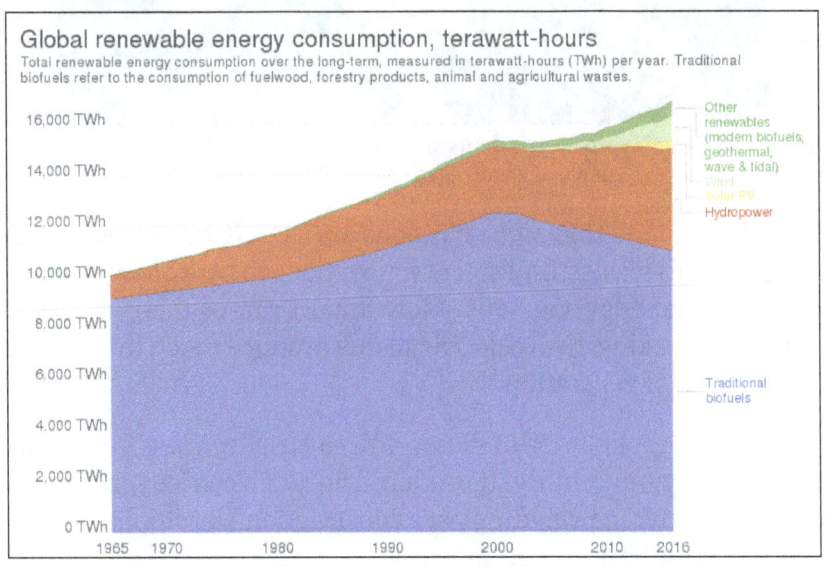

Rise in renewable energy consumption from 1965 to 2016.

When referring to sources of energy, the terms "sustainable energy" and "renewable energy" are often used interchangeably, however particular renewable energy projects sometimes raise significant sustainability concerns. Renewable energy technologies are essential contributors to sustainable energy as they generally contribute to world energy security, reducing dependence on fossil fuel resources, and providing opportunities for mitigating greenhouse gases.

Hydropower

Among sources of renewable energy, hydroelectric plants have the advantages of being long-lived—many existing plants have operated for more than 100 years. Also, hydroelectric plants are clean and have few emissions. Criticisms directed at large-scale hydroelectric plants include: dislocation of people living where the reservoirs are planned, and release of significant amounts of carbon dioxide during construction and flooding of the reservoir.

Hydroelectric dams are one of the most widely
deployed sources of sustainable energy.

However, it has been found that high emissions are associated only with shallow reservoirs in warm (tropical) locales, and recent innovations in hydropower turbine technology are enabling efficient development of low-impact run-of-the-river hydroelectricity projects. Generally speaking, hydroelectric plants produce much lower life-cycle emissions than other types of generation.

In 2015, hydropower supplied 16% of the world's electricity, down from a high of nearly 20% in the mid-to late 20th century. It produced 60% of electricity in Canada and nearly 80% Brazil. As of 2017, new hydropower construction has stopped or slowed down since 1980 in most countries except China.

Biomass

Sugarcane plantation to produce ethanol in Brazil.

A CHP power station using wood to provide electricity to over 30.000 households in France.

Biomass is biological material derived from living, or recently living organisms. As an energy source, biomass can either be burned to produce heat and to generate electricity, or converted to modern biofuels such as biodiesel and ethanol.

Biomass is extremely versatile and one of the most-used sources of renewable energy. It is available in many countries, which makes it attractive for reducing dependence on imported fossil fuels. If the production of biomass is well-managed, carbon emissions can be significantly offset by the absorption of carbon dioxide by the plants during their lifespans. If the biomass source is agricultural or municipal waste, burning it or converting it into biogas also provides a way to dispose of this waste. Bioenergy production can be combined with carbon capture and storage to create a zero-carbon or negative-carbon system.

As of 2012, wood remains the largest biomass energy source today. If biomass is harvested from crops, such as tree plantations, the cultivation of these crops can displace natural ecosystems, degrade soils, and consume water resources and synthetic fertilizers. In some cases, these impacts can actually result in higher overall carbon emissions compared to using petroleum-based fuels.

Use of farmland for growing fuel can result in less land being available for growing food. Since photosynthesis is inherently inefficient, and crops also require significant amounts of energy to harvest, dry, and transport, the amount of energy produced per unit of land area is very small, in the range of 0.25 W/m^2 to 1.2 W/m^2. In the United States, corn-based ethanol has replaced less than 10% of motor gasoline use since 2011, but has consumed around 40% of the annual corn harvest in the country. In Malaysia and Indonesia, the clearing of forests to produce palm oil for biodiesel has led to serious social and environmental effects, as these forests are critical carbon sinks and habitats for endangered species. In 2015, annual global production of liquid biofuels was equivalent to 1.8% of the energy extracted from crude oil.

Wind and Solar Electricity

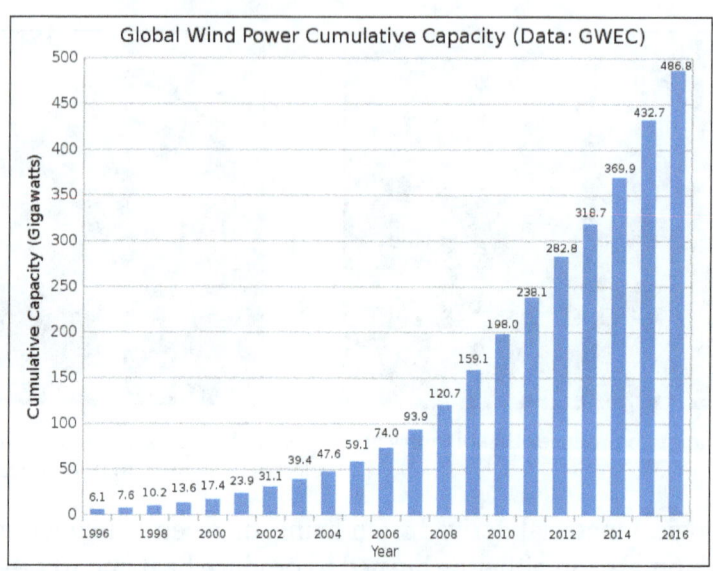

Wind power: worldwide installed capacity.

In 2015, wind power provided approximately 3.5% of the global electricity supply, and solar power provided around 1%. Wind power accounts for approximately 20% of electricity use in Denmark, 9% in Spain, and 7% in Germany. However, it may be difficult to site wind turbines in some areas for aesthetic or environmental reasons. A large wind farm may consist of several hundred individual wind turbines, and cover an extended area of hundreds of square miles, but the land between the turbines may be used for agricultural or other purposes. A wind farm may also be located offshore. A significant challenge in integrating wind energy into the electrical grid is its high intermittency. Depending on location, some land-based wind turbines generate electricity only 20% to 25% of the time; offshore wind turbines can operate up around 40% of the time.

11 MW solar power plant near Serpa, Portugal.

Solar electricity production uses photovoltaic (PV) cells to convert light into electrical current. Photovoltaic modules can be integrated into buildings or used in photovoltaic power stations connected to the electrical grid. They are especially useful for providing electricity to remote areas.

Most current solar power plants are made from an array of similar units where each unit is continuously adjusted, e.g., with some step motors, so that the light converter stays in focus of the sun light. The cost of focusing light on converters such as high-power solar panels, stirling engine, etc. can be dramatically decreased with a simple and efficient rope mechanics. In this technique many units are connected with a network of ropes so that pulling two or three ropes is sufficient to keep all light converters simultaneously in focus as the direction of the sun changes. Solar electricity is highly intermittent depending on time of day, season, and cloud cover.

Managing Intermittency for Wind and Solar Energy

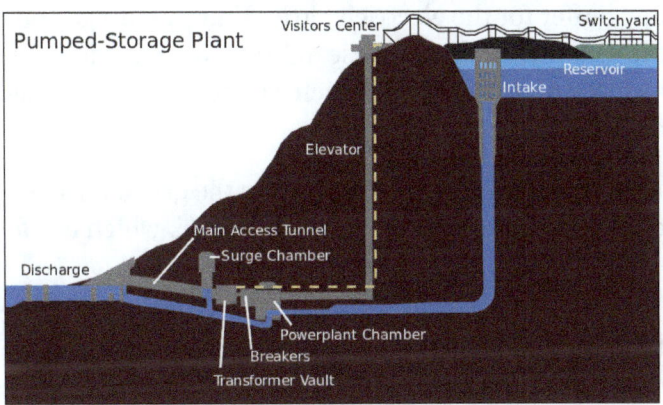

In a pumped-storage hydroelectricity facility, water is pumped uphill electricity generation exceeds demand. The water is later released to generate hydroelectricity.

Solar and wind are intermittent energy sources that supply electricity 10-40% of the time, depending on the weather and the time of day. Most electric grids were constructed for non-intermittent energy sources such as coal-fired power plants. In general, up to around 30% of the energy supplied to an electric grid can be provided by intermittent sources without requiring changes to the grid system.

If intermittent sources make up a larger percentage of the energy supply for a given electric grid, there are several possible approaches to ensuring that electricity generation can meet ongoing demand:

- Using grid energy storage to store excess solar and wind energy and release it as needed. The most commonly-used storage method is pumped-storage hydroelectricity, which is feasible only at locations that are next to a large hill or a deep underground mine. Batteries are being deployed widely. Other storage technologies are used in limited situations.

- Using hydroelectricity or natural gas generation to produce backup power.

- Importing electricity from other locations through long-distance transmission lines, such as distributing solar power from the Sahara to Europe.

- Reducing demand for electricity at certain times through energy demand management and use of smart grids.

As of 2019, the cost and logistics of energy storage for large population centres is a significant challenge, although the cost of battery systems has plunged dramatically. For instance, a 2019 study found that for solar and wind energy to meet energy demand for a week of extreme cold in the eastern and midwest United States, energy storage capacity would have to increase from the 11 GW currently in place to 277.9 GW.

Some costs could potentially be reduced by making use of energy storage equipment the consumer buys and not the state. An example is batteries in electric cars that would double as an energy buffer for the electricity grid. Energy storage apparatus' as car batteries are also built with materials that pose a threat to the environment (e.g. Lithium). The combined production of batteries for such a large part of the population would still have environmental concerns.

If renewable sources generate more electricity than the grid uses at a given time, the excess energy could potentially be stored as hydrogen fuel, which has found applications as a vehicle fuel.

Solar Heating

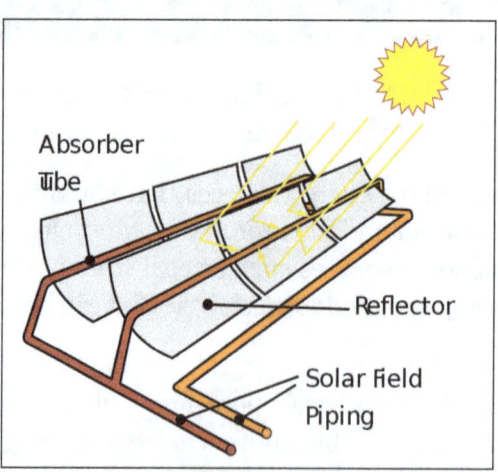

Sketch of a Parabolic Trough Collector.

Solar heating systems generally consist of solar thermal collectors, a fluid system to move the heat from the collector to its point of usage, and a reservoir or tank for heat storage and subsequent use. The systems may be used to heat domestic hot water, swimming pool water, or for space heating. The heat can also be used for industrial applications or as an energy input for other uses such as cooling equipment. In many climates, a solar heating system can provide a very high percentage (20 to 80%) of domestic hot water energy. Heat can be stored through thermal energy storage technologies. For instance, summer heat can be stored for winter heating. Similar principles are used to store winter cold for summer air conditioning.

Ocean energy refers to the energy carried by ocean waves, tides, salinity, and ocean temperature differences. The movement of water in the world's oceans creates a vast store of kinetic energy, or energy in motion. Some of this energy can be harnessed to generate electricity to power homes, transport and industries. Ocean energy has the potential of providing a substantial amount of new renewable energy around the world. As of 2017, a few small tidal power plants are operating in France and China.

One of many power plants at The Geysers, a geothermal power
field in northern California, with a total output of over 750 MW.

Geothermal energy can be harnessed to for electricity generation and for heating. Technologies in use include dry steam power stations, flash steam power stations and binary cycle power stations. As of 2010, geothermal electricity generation is used in 24 countries, while geothermal heating is in use in 70 countries. International markets grew at an average annual rate of 5 percent over the three years to 2015, and global geothermal power capacity is expected to reach 14.5–17.6 GW by 2020.

Geothermal power is considered to be a sustainable, renewable source of energy because the heat extraction is small compared with the Earth's heat content. The greenhouse gas emissions of geothermal electric stations are on average 45 grams of carbon dioxide per kilowatt-hour of electricity, or less than 5 percent of that of conventional coal-fired plants. As a source of renewable energy for both power and heating, geothermal has the potential to meet 3-5% of global demand by 2050. With economic incentives, it is estimated that by 2100 it will be possible to meet 10% of global demand.

Non-renewable Energy Sources

Nuclear Power

Nuclear power plants have been used since the 1950s to produce a steady supply of electricity, without creating local air pollution. In 2012, nuclear power plants in 30 countries generated 11% of global electricity. The IPCC considers nuclear power to be a low-carbon energy source, with lifecycle greenhouse gas emissions (including the mining and processing of uranium), similar to the emissions from renewable energy sources.

There is considerable controversy over whether nuclear power can be considered sustainable, with debates revolving around the risk of nuclear accidents, the cost and construction time needed to build new plants, the generation of radioactive nuclear waste, and the potential for nuclear energy to contribute to nuclear proliferation. These concerns have led to a decrease in the contribution of nuclear energy to the global electricity supply since 1993. At a global level, opposition to nuclear energy stood at 62 percent in 2011. Public support for nuclear energy is often low as a result of safety concerns, however for each unit of energy produced, nuclear energy is far safer than fossil fuel energy.

Traditional environmental groups such as Greenpeace and the Sierra Club are opposed to all use of nuclear power. Individuals who have described nuclear power as a green energy source include early Greenpeace member Patrick Moore, Stewart Brand, George Monbiot, Bill Gates and James Lovelock.

Newer nuclear reactor designs are capable of burning nuclear waste until it is no longer (or dramatically less) dangerous, and have design features that greatly minimize the possibility of a nuclear accident. These designs have yet to be commercialized. Some forms of nuclear power can "burn" nuclear waste through a process known as nuclear transmutation, such as an Integral Fast Reactor. Nuclear power plants can be more or less eliminated from their problem of nuclear waste through the use of nuclear reprocessing and newer plants such as fast breeder plants.

Fuel Switching

For a given unit of energy produced, the greenhouse gas emissions of natural gas are around half the emissions of coal when used to generate electricity, and around two-thirds the emissions of coal when used to produce heat. Natural gas also produces significantly less air pollution than coal. Building gas-fired power plants and gas pipelines is therefore promoted as a way to reduce emissions and phase out coal use, however this practice is controversial. Opponents argue that developing natural gas infrastructure will create decades of technology lock-in for fossil fuels, and that renewables create far less emissions at comparable costs. The life-cycle greenhouse-gas emissions of natural gas are around 40 times the emissions of wind energy.

Carbon Capture and Storage

In theory, the greenhouse gas emissions of fossil fuel and biomass power plants can be significantly reduced through carbon capture and storage, although this process is expensive.

Energy Efficiency

Moving towards energy sustainability will require changes not only in the way energy is supplied, but in the way it is used, and reducing the amount of energy required to deliver various goods or services is essential. Opportunities for improvement on the demand side of the energy equation are as rich and diverse as those on the supply side, and often offer significant economic benefits.

Efficiency slows down energy demand growth so that rising clean energy supplies can make deep cuts in fossil fuel use. A recent historical analysis has demonstrated that the rate of energy efficiency improvements has generally been outpaced by the rate of growth in energy demand, which is due to continuing economic and population growth. As a result, despite energy efficiency gains, total energy use and related carbon emissions have continued to increase. Thus, given the thermodynamic and practical limits of energy efficiency improvements, slowing the growth in energy demand is essential. However, unless clean energy supplies come online rapidly, slowing demand growth will only begin to reduce total emissions; reducing the carbon content of energy sources is also needed. Any serious vision of a sustainable energy economy thus requires commitments to both renewables and efficiency.

Trends

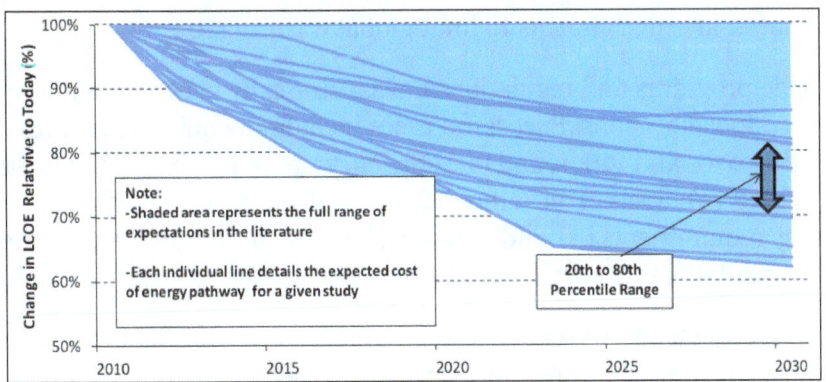

The National Renewable Energy Laboratory projects that the levelized cost of wind power in the U.S. will decline about 25% from 2012 to 2030.

Climate change concerns coupled with high oil prices and increasing government support are driving increasing rates of investment in the sustainable energy industries, according to a trend analysis from the United Nations Environment Programme. According to UNEP, global investment in sustainable energy in 2007 was higher than previous levels, with $148 billion of new money raised in 2007, an increase of 60% over 2006. Total financial transactions in sustainable energy, including acquisition activity, was $204 billion.

Investment flows in 2007 broadened and diversified, making the overall picture one of greater breadth and depth of sustainable energy use. The mainstream capital markets are "now fully receptive to sustainable energy companies, supported by a surge in funds destined for clean energy investment". The increased levels of investment and the fact that much of the capital is coming from more conventional financial actors suggest that sustainable energy options are now becoming mainstream.

Government Promotion of Sustainable Energy

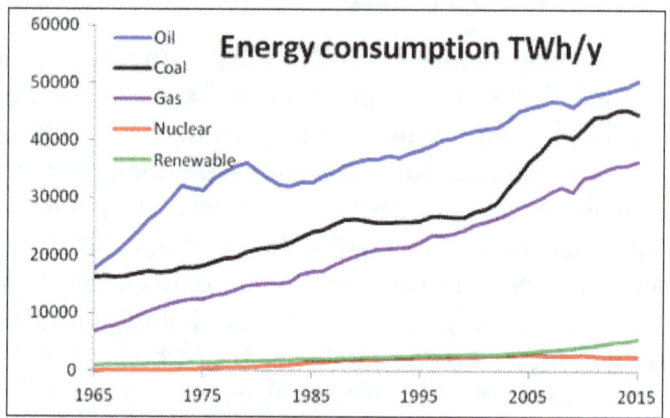

Comparing trends in worldwide energy use, the growth of renewable energy to 2015 is shown by the green line.

According to the IPCC, both explicit carbon pricing and complementary energy-specific policies are necessary mechanisms to limit global warming to 1.5°C.

Energy-specific programs and regulations have historically been the mainstay of efforts to reduce fossil fuel emissions. Successful cases include the building of nuclear reactors in France in the 1970s and 1980s, and fuel efficiency standards in the United States which conserved billions of barrels of oil. Other examples of energy-specific policies include energy-efficiency requirements in building codes, banning new coal-fired electricity plants, performance standards for electrical appliances, and support for electric vehicle use.

Carbon taxes are an effective way to encourage movement towards a low-carbon economy, while providing a source of revenue that can be used to keep energy affordable. Carbon taxes have encountered strong political pushback in some jurisdictions, whereas energy-specific policies tend to be politically safer. As of 2018, carbon prices are much lower than the actual costs of emissions to society: In 42 OECD countries and territories they averaged USD$8 per ton of carbon dioxide emitted, around a quarter of the OECD's estimated actual cost of carbon. Some studies estimate that combining a carbon tax with energy-specific policies would be more cost-effective than a carbon tax alone.

Sustainable Energy Research

There are numerous organizations within the academic, federal, and commercial sectors conducting large scale advanced research in the field of sustainable energy. Scientific production towards sustainable energy systems is rising exponentially, growing from about 500 English journal papers only about renewable energy in 1992 to almost 9,000 papers in 2011.

Biofuels

Cellulosic ethanol has many benefits over traditional corn based-ethanol. It does not take away or directly conflict with the food supply because it is produced from wood, grasses, or non-edible parts of plants. Moreover, some studies have shown cellulosic ethanol to be potentially more cost effective and economically sustainable than corn-based ethanol. As of 2018, efforts to commercialize production of cellulosic ethanol have been mostly disappointing, but new commercial efforts are continuing.

Algae fuel is an alternative to liquid fossil fuels that uses algae as its source of energy-rich oils. During the biofuel production process algae actually consumes the carbon dioxide in the air and turns it into oxygen through photosynthesis. In addition to its projected high yield, algaculture— unlike food crop-based biofuels — does not entail a decrease in food production, since it requires neither farmland nor fresh water. Between 2005 and 2012, dozens of companies attempted to commercialize production of algae fuel. By 2017, however, most efforts had been abandoned or changed to other applications, with only a few remaining.

Nuclear

There are potentially two sources of nuclear power. Fission is used in all current nuclear power plants. Fusion is the reaction that exists in stars, including the sun, and remains impractical for use on Earth, as fusion reactors are not yet available. However nuclear power is controversial politically and scientifically due to concerns about radioactive waste disposal, safety, the risks of a severe accident, and technical and economical problems in dismantling of old power plants.

Thorium is a fissionable material used in thorium-based nuclear power. The thorium fuel cycle claims several potential advantages over a uranium fuel cycle, including greater abundance, superior physical and nuclear properties, better resistance to nuclear weapons proliferation and reduced plutonium and actinide production. Therefore, it is sometimes referred as sustainable.

Solar

Currently, photovoltaic (PV) panels only have the ability to convert around 24% of the sunlight that hits them into electricity. At this rate, solar energy still holds many challenges for widespread implementation, but steady progress has been made in reducing manufacturing cost and increasing photovoltaic efficiency. In 2008, researchers at Massachusetts Institute of Technology (MIT) developed a method to store solar energy by using it to produce hydrogen fuel from water. Such research is targeted at addressing the obstacle that solar development faces of storing energy for use during nighttime hours when the sun is not shining. In February 2012, North Carolina-based Semprius Inc., announced that they had developed the world's most efficient solar panel. The company claims that the prototype converts 33.9% of the sunlight that hits it to electricity, more than double the previous high-end conversion rate. Major projects on artificial photosynthesis or solar fuels are also under way in many developed nations.

Large national and regional research projects on artificial photosynthesis are designing nanotechnology-based systems that use solar energy to split water into hydrogen fuel. and a proposal has been made for a Global Artificial Photosynthesis project In 2011, researchers at the Massachusetts Institute of Technology (MIT) developed what they are calling an "Artificial Leaf", which is capable of splitting water into hydrogen and oxygen directly from solar power when dropped into a glass of water. One side of the "Artificial Leaf" produces bubbles of hydrogen, while the other side produces bubbles of oxygen.

Research is ongoing in space-based solar power, a concept in which solar panels are launched into outer space and the energy they capture is transmitted back to Earth as microwaves. A test facility for the technology is being built in China.

MIT's Solar House 1 built in 1939 used seasonal thermal energy storage (STES) for year-round heating.

Geothermal

Geothermal energy is produced by tapping into the thermal energy created and stored within the earth. It arises from the radioactive decay of an isotope of potassium and other elements found in the Earth's crust. Geothermal energy can be obtained by drilling into the ground, very similar to oil exploration, and then it is carried by a heat-transfer fluid (e.g. water, brine or steam). Geothermal systems that are mainly dominated by water have the potential to provide greater benefits to the system and will generate more power. Within these liquid-dominated systems, there are possible concerns of subsidence and contamination of ground-water resources. Therefore, protection of ground-water resources is necessary in these systems. This means that careful reservoir production and engineering is necessary in liquid-dominated geothermal reservoir systems. Geothermal energy is considered sustainable because that thermal energy is constantly replenished.

Hydrogen

Over $1 billion of federal money has been spent on the research and development of hydrogen and a medium for energy storage in the United States. Hydrogen is useful for energy storage, and for use in airplanes and ships, but is not practical for automobile use, as it is not very efficient, compared to using a battery — for the same cost a person can travel three times as far using a battery electric vehicle. Regardless of that opinion, Japanese car manufacturers Toyota and Honda currently offer hydrogen fuel-cell powered passenger vehicles for sale in Japan and the U.S.A. Experimental hydrogen fuel-cell city buses are currently operative in two U.S. transit districts, Alameda/Contra Costa county, California, and in Connecticut.

SUSTAINABLE TRANSPORT

Possible scenario of clean mobility.

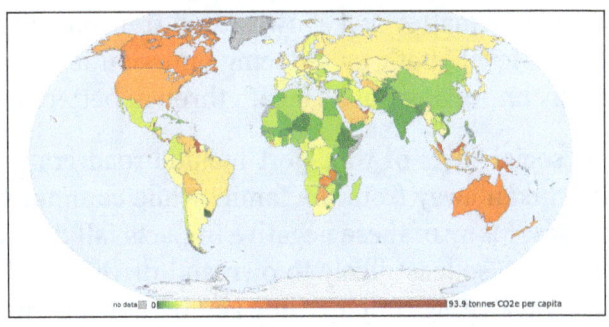

Anthropogenic per capita emissions of greenhouse gases by country by the year 2000.

Sustainable transport refers to the broad subject of transport that is sustainable in the senses of social, environmental and climate impacts. Components for evaluating sustainability include the particular vehicles used for road, water or air transport; the source of energy; and the infrastructure used to accommodate the transport (roads, railways, airways, waterways, canals and terminals). Transport operations and logistics as well as transit-oriented development are also involved in evaluation. Transportation sustainability is largely being measured by transportation system effectiveness and efficiency as well as the environmental and climate impacts of the system.

Short-term activity often promotes incremental improvement in fuel efficiency and vehicle emissions controls while long-term goals include migrating transportation from fossil-based energy to other alternatives such as renewable energy and use of other renewable resources. The entire life cycle of transport systems is subject to sustainability measurement and optimization.

Sustainable transport systems make a positive contribution to the environmental, social and economic sustainability of the communities they serve. Transport systems exist to provide social and economic connections, and people quickly take up the opportunities offered by increased mobility, with poor households benefiting greatly from low carbon transport options. The advantages of increased mobility need to be weighed against the environmental, social and economic costs that transport systems pose.

Transport systems have significant impacts on the environment, accounting for between 20% and 25% of world energy consumption and carbon dioxide emissions. The majority of the emissions, almost 97%, came from direct burning of fossil fuels. Greenhouse gas emissions from transport are increasing at a faster rate than any other energy using sector. Road transport is also a major contributor to local air pollution and smog.

The United Nations Environment Programme (UNEP) estimates that each year 2.4 million premature deaths from outdoor air pollution could be avoided. Particularly hazardous for health are emissions of black carbon, a component of particulate matter, which is a known cause of respiratory and carcinogenic diseases and a significant contributor to global climate change. The links between greenhouse gas emissions and particulate matter make low carbon transport an increasingly sustainable investment at local level—both by reducing emission levels and thus mitigating climate change; and by improving public health through better air quality.

The social costs of transport include road crashes, air pollution, physical inactivity, time taken away from the family while commuting and vulnerability to fuel price increases. Many of these negative impacts fall disproportionately on those social groups who are also least likely to own and drive cars. Traffic congestion imposes economic costs by wasting people's time and by slowing the delivery of goods and services.

Traditional transport planning aims to improve mobility, especially for vehicles, and may fail to adequately consider wider impacts. But the real purpose of transport is access – to work, education, goods and services, friends and family – and there are proven techniques to improve access while simultaneously reducing environmental and social impacts, and managing traffic congestion. Communities which are successfully improving the sustainability of their transport networks are doing so as part of a wider programme of creating more vibrant, livable, sustainable cities.

The term sustainable transport came into use as a logical follow-on from sustainable development, and is used to describe modes of transport, and systems of transport planning, which are consistent with wider concerns of sustainability. There are many definitions of the sustainable transport, and of the related terms sustainable transportation and sustainable mobility. One such definition, from the European Union Council of Ministers of Transport, defines a sustainable transportation system as one that:

- Allows the basic access and development needs of individuals, companies and society to be met safely and in a manner consistent with human and ecosystem health, and promotes equity within and between successive generations.

- Is affordable, operates fairly and efficiently, offers a choice of transport mode, and supports a competitive economy, as well as balanced regional development.

- Limits emissions and waste within the planet's ability to absorb them, uses renewable resources at or below their rates of generation, and uses non-renewable resources at or below the rates of development of renewable substitutes, while minimizing the impact on the use of land and the generation of noise.

Sustainability extends beyond just the operating efficiency and emissions. A life-cycle assessment involves production, use and post-use considerations. A cradle-to-cradle design is more important than a focus on a single factor such as energy efficiency.

Environmental Impact

The Bus Rapid Transit of Metz uses a diesel-electric hybrid driving system, developed by Belgian Van Hool manufacturer.

Electric Transmetro in Guatemala City.

Transport systems are major emitters of greenhouse gases, responsible for 23% of world energy-related GHG emissions in 2004, with about three quarters coming from road vehicles. Currently 95% of transport energy comes from petroleum. Energy is consumed in the manufacture as well as the use of vehicles, and is embodied in transport infrastructure including roads, bridges and railways.

The first historical attempts of evaluating the Life Cycle environmental impact of vehicle is due to Theodore Von Karman. After decades in which all the analysis has been focused on emending the Von Karman model, Dewulf and Van Langenhove have introduced an model based on the second law of thermodynamics and exergy analysis. Chester and Orwath, have developed a similar model based on the first law that accounts the necessary costs for the infrastructure.

The environmental impacts of transport can be reduced by reducing the weight of vehicles, sustainable styles of driving, reducing the friction of tires, encouraging electric and hybrid vehicles, improving the walking and cycling environment in cities, and by enhancing the role of public transport, especially electric rail.

Green vehicles are intended to have less environmental impact than equivalent standard vehicles, although when the environmental impact of a vehicle is assessed over the whole of its life cycle this may not be the case.

Electric vehicle technology (especially non-battery based vehicles, fuel cell vehicles) has the potential to reduce transport CO_2 emissions, depending on the embodied energy of the vehicle and the source of the electricity. The primary sources of electricity currently used in most countries (coal, gas, oil) mean that until world electricity production changes substantially, private electric cars will result in the same or higher production of CO_2 than petrol equivalent vehicles. Battery-based electric vehicles may or may not be better in terms of GHG emissions then fossil-fuel based vehicles depending on several factors, such as battery type, capacity of the battery, life expectancy of the battery.

The Online Electric Vehicle (OLEV), developed by the Korea Advanced Institute of Science and Technology (KAIST), is an electric vehicle that can be charged while stationary or driving, thus removing the need to stop at a charging station. The City of Gumi in South Korea runs a 24 km roundtrip along which the bus will receive 100 kW (136 horsepower) electricity at an 85% maximum power transmission efficiency rate while maintaining a 17 cm air gap between the underbody of the vehicle and the road surface. At that power, only a few sections of the road need embedded cables. Hybrid vehicles, which use an internal combustion engine combined with an electric engine to achieve better fuel efficiency than a regular combustion engine, are already common.

Natural gas is also used as a transport fuel but is a less promising, technology as it is still a fossil fuel and still has significant emissions (though lower then gasoline, diesel).

Brazil met 17% of its transport fuel needs from bioethanol in 2007, but the OECD has warned that the success of (first-generation) biofuels in Brazil is due to specific local circumstances. Internationally, first-generation biofuels are forecast to have little or no impact on greenhouse emissions, at significantly higher cost than energy efficiency measures. The later generation biofuels however (2nd to 4th generation) do have significant environmental benefit, as they are no driving force for deforestation or struggle with the food vs fuel issue. Other renewable fuels include hydrogen, which (like drop-in biofuels) can be used in internal combustion vehicles, don't rely on any crops at all (instead it's produced using electricity) and even generates very little pollution when burned.

In practice there is a sliding scale of green transport depending on the sustainability of the option. Green vehicles are more fuel-efficient, but only in comparison with standard vehicles, and they still contribute to traffic congestion and road crashes. Well-patronised public transport networks based on traditional diesel buses use less fuel per passenger than private vehicles, and are generally safer and use less road space than private vehicles. Green public transport vehicles including electric trains, trams and electric buses combine the advantages of green vehicles with those of sustainable transport choices. Other transport choices with very low environmental impact are cycling and other human-powered vehicles, and animal powered transport. The most common green transport choice, with the least environmental impact is walking.

Transport on Rails Boasts an Excellent Efficiency

Transport and Social Sustainability

A tram in Melbourne.

Cities with overbuilt roadways have experienced unintended consequences, linked to radical drops in public transport, walking, and cycling. In many cases, streets became void of "life." Stores, schools, government centers and libraries moved away from central cities, and residents who did not flee to the suburbs experienced a much reduced quality of public space and of public services. As schools were closed their mega-school replacements in outlying areas generated additional traffic; the number of cars on US roads between 7:15 and 8:15 a.m. increases 30% during the school year.

Yet another impact was an increase in sedentary lifestyles, causing and complicating a national epidemic of obesity, and accompanying dramatically increased health care costs.

Cities

Futurama, an exhibit at the 1939 New York World's Fair, was sponsored by General Motors and showed a vision of the City of Tomorrow.

Cities are shaped by their transport systems. In The City in History, Lewis Mumford documented how the location and layout of cities was shaped around a walkable center, often located near a port or waterway, and with suburbs accessible by animal transport or, later, by rail or tram lines.

In 1939, the New York World's Fair included a model of an imagined city, built around a car-based transport system. In this "greater and better world of tomorrow", residential, commercial and industrial areas were separated, and skyscrapers loomed over a network of urban motorways. These ideas captured the popular imagination, and are credited with influencing city planning from the 1940s to the 1970s.

Interstate 10 and Interstate 45 near downtown Houston, Texas.

The popularity of the car in the post-war era led to major changes in the structure and function of cities. There was some opposition to these changes at the time. The writings of Jane Jacobs, in particular The Death and Life of Great American Cities provide a poignant reminder of what was lost in this transformation, and a record of community efforts to resist these changes. Lewis Mumford asked "is the city for cars or for people?" Donald Appleyard documented the consequences for communities of increasing car traffic in "The View from the Road" and in the UK, Mayer Hillman first published research into the impacts of traffic on child independent mobility in 1971. Despite these notes of caution, trends in car ownership, car use and fuel consumption continued steeply upward throughout the post-war period.

Mainstream transport planning in Europe has, by contrast, never been based on assumptions that the private car was the best or only solution for urban mobility. For example, the Dutch Transport Structure Scheme has since the 1970s required that demand for additional vehicle capacity only be met "if the contribution to societal welfare is positive", and since 1990 has included an explicit target to halve the rate of growth in vehicle traffic. Some cities outside Europe have also consistently linked transport to sustainability and to land-use planning, notably Curitiba, Brazil, Portland, Oregon and Vancouver, Canada.

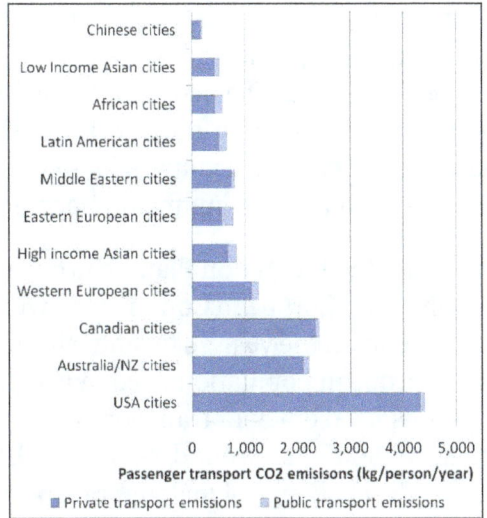

Greenhouse gas emissions from transport vary widely,
even for cities of comparable wealth.

There are major differences in transport energy consumption between cities; an average U.S. urban dweller uses 24 times more energy annually for private transport than a Chinese urban resident, and almost four times as much as a European urban dweller. These differences cannot be explained by wealth alone but are closely linked to the rates of walking, cycling, and public transport use and to enduring features of the city including urban density and urban design.

A bypass the Old Town in Szczecin, Poland.

The cities and nations that have invested most heavily in car-based transport systems are now the least environmentally sustainable, as measured by per capita fossil fuel use. The social and economic sustainability of car-based transportation engineering has also been questioned. Within the United States, residents of sprawling cities make more frequent and longer car trips, while residents of traditional urban neighbourhoods make a similar number of trips, but travel shorter distances and walk, cycle and use transit more often. It has been calculated that New York residents save $19 billion each year simply by owning fewer cars and driving less than the average American. A less car intensive means of urban transport is carsharing, which is becoming popular in North America and Europe, and according to The Economist, carsharing can reduce car ownership at an estimated rate of one rental car replacing 15 owned vehicles. Car sharing has also begun in the developing world, where traffic and urban density is often worse than in developed countries. Companies like Zoom in India, eHi in China, and Carrot in Mexico, are bringing car-sharing to developing countries in an effort to reduce car-related pollution, ameliorate traffic, and expand the number of people who have access to cars.

The European Commission adopted the Action Plan on urban mobility on 2009-09-30 for sustainable urban mobility. The European Commission will conduct a review of the implementation of the Action Plan in the year 2012, and will assess the need for further action. In 2007, 72% of the European population lived in urban areas, which are key to growth and employment. Cities need efficient transport systems to support their economy and the welfare of their inhabitants. Around 85% of the EU's GDP is generated in cities. Urban areas face today the challenge of making transport sustainable in environmental (CO_2, air pollution, noise) and competitiveness (congestion) terms while at the same time addressing social concerns. These range from the need to respond to health problems and demographic trends, fostering economic and social cohesion to taking into account the needs of persons with reduced mobility, families and children.

Policies and Governance

Sustainable transport policies have their greatest impact at the city level. Outside Western Europe, cities which have consistently included sustainability as a key consideration in transport and land use planning include Curitiba, Brazil; Bogota, Colombia; Portland, Oregon; and Vancouver, Canada. The state of Victoria, Australia passed legislation in 2010 – the Transport Integration Act – to compel its transport agencies to actively consider sustainability issues including climate change impacts in transport policy, planning and operations.

Many other cities throughout the world have recognised the need to link sustainability and transport policies, for example by joining the Cities for Climate Protection program.

Seven sustainable transportations in one photo (Prague).

Community and Grassroots Action

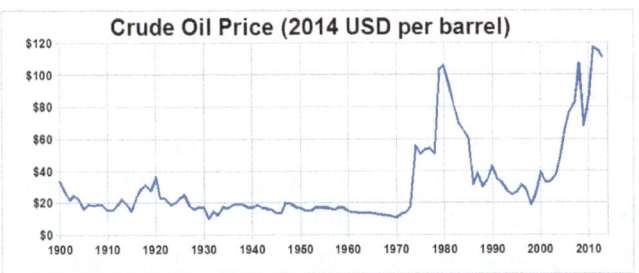

Oil price trend, 1939–2007, both nominal and adjusted to inflation.

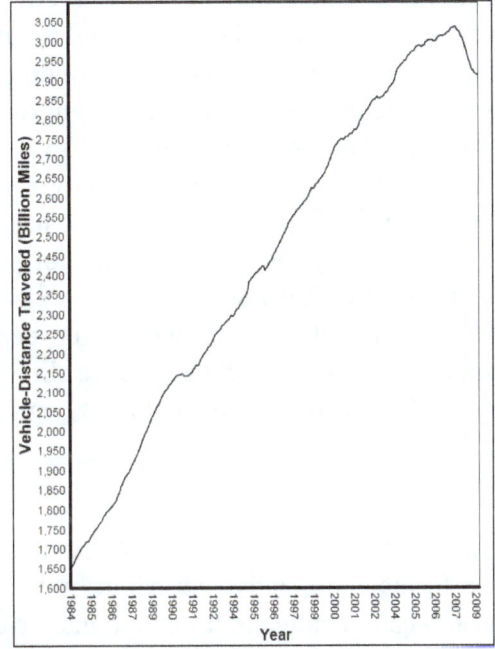

Vehicle-miles traveled in the United States up to March 2009.

Sustainable transport is fundamentally a grassroots movement, albeit one which is now recognised as of citywide, national and international significance.

Whereas it started as a movement driven by environmental concerns, over these last years there has been increased emphasis on social equity and fairness issues, and in particular the need to ensure proper access and services for lower income groups and people with mobility limitations, including the fast-growing population of older citizens. Many of the people exposed to the most vehicle noise, pollution and safety risk have been those who do not own, or cannot drive cars, and those for whom the cost of car ownership causes a severe financial burden.

An organization called Greenxc started in 2011 created a national awareness campaign in the United States encouraging people to carpool by ride-sharing cross country stopping over at various destinations along the way and documenting their travel through video footage, posts and photography. Ride-sharing reduces individual's carbon footprint by allowing several people to use one car instead of everyone using individual cars.

Tools and Incentives

Several European countries are opening up financial incentives that support more sustainable modes of transport. The European Cyclists' Federation, which focuses on daily cycling for transport, has created a document containing a non-complete overview. In the UK, employers have for many years been providing employees with financial incentives. The employee leases or borrows a bike that the employer has purchased. You can also get other support. The scheme is beneficial for the employee who saves money and gets an incentive to get exercise integrated in the daily routine. The employer can expect a tax deduction, lower sick leave and less pressure on parking spaces for cars. Since 2010, there has been a scheme in Iceland (Samgöngugreiðslur) where those who do not drive a car to work, get paid a lump of money monthly. An employee must sign a statement not to use a car for work more often than one day a week, or 20% of the days for a period. Some employers pay fixed amounts based on trust. Other employers reimburse the expenses for repairs on bicycles, period-tickets for public transport and the like. Since 2013, amounts up to ISK 8000 per month have been tax-free. Most major workplaces offer this, and a significant proportion of employees use the scheme. From the year 2019 half the amount is tax-free if the employee sings a contract not to use a car to work for more than 40% of the days of the contract period.

Greenwashing

The term green transport is often used as a greenwash marketing technique for products which are not proven to make a positive contribution to environmental sustainability. Such claims can be legally challenged. For instance Norway's consumer ombudsman has targeted automakers who claim that their cars are "green", "clean" or "environmentally friendly". Manufacturers risk fines if they fail to drop the words.

CARBON DIOXIDE REMOVAL

Carbon dioxide removal (CDR) refers to a group of technologies the objective of which is the large-scale removal of carbon dioxide from the atmosphere. Among such technologies are bio-energy with carbon capture and storage, biochar, ocean fertilization, enhanced weathering, and direct air capture when combined with storage. CDR is a different approach from removing CO_2 from the stack emissions of large fossil fuel point sources, such as power stations. The latter reduces emission to the atmosphere but cannot reduce the amount of carbon dioxide already in the atmosphere. As CDR removes carbon dioxide from the atmosphere, it 'creates' negative emissions that offset the emissions from small and dispersed point sources such as domestic heating systems, airplanes and vehicle exhausts. It is regarded by some as a form of climate engineering, while other commentators describe it as a form of carbon capture and storage or extreme mitigation. Whether CDR would satisfy common definitions of "climate engineering" or "geoengineering" usually depends upon the scale at which it would be undertaken.

The likely need for CDR has been publicly expressed by a range of individuals and organizations involved with climate change issues.

The mitigation effectiveness of air capture is limited by societal investment, land use, availability of geologic reservoirs, and leakage. The reservoirs are estimated to be sufficient for storing at least 545 gigatonnes of carbon (GtC). Storing 771 GtC would cause a 186 ppm atmospheric reduction. In order to return the atmospheric CO_2 content to 350 ppm we would need atmospheric reductions of 50 ppm each year and also to reduce current emissions by the equivalent of 2 ppm per year.

Carbon dioxide removal is different from reducing emissions, as the former produces an outlet of carbon dioxide from Earth's atmosphere, whereas the latter decreases the inlet of carbon dioxide to the atmosphere. Both have the same net effect, but for achieving carbon dioxide concentration levels below present levels, carbon dioxide removal is critical. Also for meeting higher concentration levels, carbon dioxide removal is increasingly considered to be crucial as it provides the only possibility to fill the gap between needed reductions to meet mitigation targets and global emission trends.

In the OECD Environmental Outlook to 2050 released at the 2011 United Nations Climate Change Conference, the authors commented on the need for negative emissions, stating "Achieving lower concentration targets (450 ppm) depends significantly on the use of BECCS".

A carbon dioxide sink such as a concentrated group of plants or any other primary producer that binds carbon dioxide into biomass, such as within forests and kelp beds, is not carbon negative, as sinks are not permanent. A carbon dioxide sink of this type moves carbon, in the form of carbon dioxide, from the atmosphere or hydrosphere to the biosphere. This process could be undone, for example by wildfires or logging.

Carbon dioxide sinks that store carbon dioxide in the Earth's crust by injecting it into the subsurface, or in the form of insoluble carbonate salts (mineral sequestration), are considered carbon negative. This is because they are removing carbon from the atmosphere and sequestering it indefinitely and presumably for a considerable duration (thousands to millions of years). However, Carbon Capture technology remains, at best, theoretical and is yet to reach more than 33% efficiency. Furthermore, this process could be rapidly undone, for example by earthquakes or mining.

Methods

Bio-energy with Carbon Capture and Storage

Bio-energy with carbon capture and storage, or BECCS, uses biomass to extract carbon dioxide from the atmosphere, and carbon capture and storage technologies to concentrate and permanently store it in deep geological formations.

BECCS is currently as of October 2012 the only CDR technology deployed at full industrial scale, with 550 000 tonnes CO_2/year in total capacity operating, divided between three different facilities as of January 2012.

The Imperial College London, the UK Met Office Hadley Centre for Climate Prediction and Research, the Tyndall Centre for Climate Change Research, the Walker Institute for Climate System Research, and the Grantham Institute for Climate Change issued a joint report on carbon dioxide removal technologies as part of the AVOID: Avoiding dangerous climate change research program, stating that "Overall, of the technologies studied in this report, BECCS has the greatest maturity and there are no major practical barriers to its introduction into today's energy system. The presence of a primary product will support early deployment."

According to the OECD, "Achieving lower concentration targets (450 ppm) depends significantly on the use of BECCS".

Biochar

Biochar is created by the pyrolysis of biomass, and is under investigation as a method of carbon sequestration. Biochar is a charcoal that is used for agricultural purposes which also aids in carbon sequestration, the capture or hold of carbon. It is created using a process called pyrolysis, which is basically the act of high temperature heating biomass in an environment with low oxygen levels. What remains is a material known as char, similar to charcoal but is made through a sustainable process, thus the use of biomass. Biomass is organic matter produced by living organisms or recently living organisms, most commonly plants or plant based material. The offset of greenhouse gas (GHG) emission, if biochar were to be implemented, would be a maximum of 12%. This equates to about 106 metric tons of CO_2 equivalents. On a medium conservative level, it would be 23% less than that, at 82 metric tons. A study done by the UK Biochar Research Center has stated that, on a conservative level, biochar can store 1 gigaton of carbon per year. With greater effort in marketing and acceptance of biochar, the benefit could be the storage of 5–9 gigatons per year of carbon in biochar soils.

Enhanced Weathering

Enhanced weathering is a chemical approach to remove carbon dioxide involving land- or ocean-based techniques. One example of a land-based enhanced weathering technique is in-situ carbonation of silicates. Ultramafic rock, for example, has the potential to store from hundreds to thousands of years' worth of CO_2 emissions, according to estimates. Ocean-based techniques involve alkalinity enhancement, such as grinding, dispersing, and dissolving olivine, limestone, silicates, or calcium hydroxide to address ocean acidification and CO_2 sequestration. Enhanced weathering is considered one of the least expensive geoengineering options. One example of a research project on the feasibility of enhanced weathering is the CarbFix project in Iceland.

Direct Air Capture (DAC)

Carbon dioxide can be removed from ambient air through chemical processes, sequestered, and stored. Traditional modes of carbon capture such as precombustion and postcombustion CO_2 capture from large point sources can help slow the rate of increase of the atmospheric CO_2 concentration, but only the direct removal of CO_2 from the air, or direct air capture (DAC), can actually reduce the global atmospheric CO_2 concentration if combined with long-term storage of CO2.

DAC relying on amine-based absorption demands significant water input. It was estimated, that to capture 3.3 Gigatonnes of CO_2 a year would require 300 km³ of water, or 4% of the water used for irrigation. On the other hand, using sodium hydroxide needs far less water, but the substance itself is highly caustic and dangerous.

DAC also requires much greater energy input in comparison to traditional capture from point sources, like flue gas, due to the low concentration of CO_2. The theoretical minimum energy required to extract CO_2 from ambient air is about 250 kWh per tonne of CO_2, while capture from natural gas and coal power plants requires respectively about 100 and 65 kWh per tonne of CO_2.

Ocean Fertilization

Ocean fertilization or ocean nourishment is a type of climate engineering based on the purposeful introduction of nutrients to the upper ocean to increase marine food production and to remove carbon dioxide from the atmosphere. A number of techniques, including fertilization by iron, urea and phosphorus have been proposed.

Wooden Building Construction

In their "roadmap for rapid decarbonization", Rockström et al. proposed in the journal Science that "after 2030, all building construction must be carbon-neutral or carbon-negative. The construction industry must either use emissions-free concrete and steel or replace those materials with zero- or negative-emissions substances such as wood, stone, and carbon fiber but not tools of daily use. " Especially wooden building constructions are a promising way to store emissions, as wood mainly contains carbon, grows faster due to the higher CO_2 concentration and is established as a building material for several centuries. Projects like W350 in Tokyo and Mjøstårnet in Brumunddal demonstrate that even skyscrapers could potentially be built from wood and, thus, store large amounts of carbon.

Example CO_2 Scrubbing Chemistry

Calcium Oxide

Calcium oxide (quicklime) will absorb CO_2 from atmospheric air mixed with steam at 400 °C (forming calcium carbonate) and release it at 1,000 °C. This process, proposed by A. Steinfeld, can be performed using renewable energy from thermal concentrated solar power. Quicklime is made by heating limestone to release the CO_2 within it. Quicklime is mixed with sand for brick building as mortar, where it hardens by absorption of CO_2.

Sodium Hydroxide

Zeman and Lackner outlined a specific method of air capture using sodium hydroxide. Carbon Engineering, a Calgary, Alberta firm founded in 2009 and partially funded by Bill Gates, is developing a process to capture carbon dioxide using a solution of potassium hydroxide mixed with some water at their pilot plant. They hope to create and sell synthetic fuels at a cost of $100 a ton. They have partnered with Greyrock.

Direct Air Capture (DAC) Process Example using NaOH

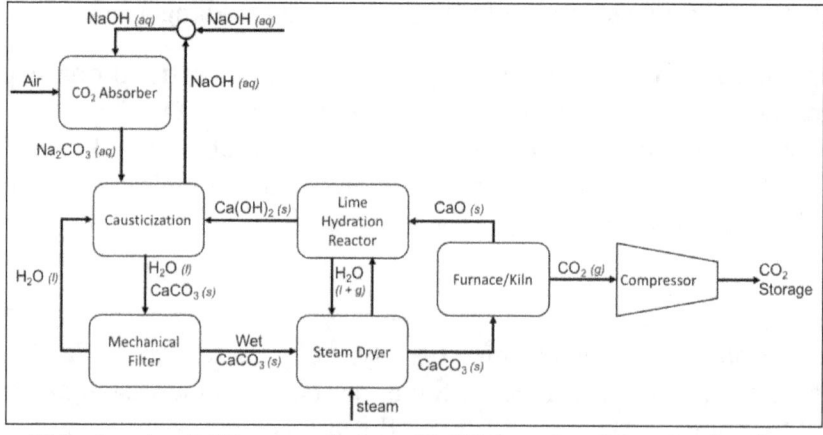

Main steps in a DCA process that uses NaOH (caustic soda) as the absorber.

Among the technologies studied for direct air capture (DAC), the use of aqueous hydroxide sorbents is one of the most promising approaches. In this process, CO_2 from the air is chemically dissolved into NaOH(aq) solution as Na_2CO_3; the Na_2CO_3 is then reacted with solid $Ca(OH)_2$, which regenerates the solvent and produces $CaCO_3$ crystals; lastly, heat is applied to the $CaCO_3$ crystals to produce pure CO_2 gas.

Air is pumped through the CO_2 absorber as the first step of this process. CO_2 absorber for DAC are designed either as a counter-current spray tower or as a counter-current thin-falling-film contractor to maximize the contact area between the air and the solvent and thus maximize the absorption driving force. The solvent is regenerated in the causticization unit by reacting the Na_2CO_3 with $Ca(OH)_2$, which also transfers the captured CO_2 to the form of $CaCO_3$ solid crystals. A mechanical filter is then used to separate the $CaCO_3$ crystals from the water. Since the crystals come out wet from the filter, they are dried in a steam dryer. Then the dry crystals are heated in a furnace to produce CaO and pure CO_2 gas. The CaO is then hydrated to regenerate the $Ca(OH)_2$ used for the causticization reaction. The pure CO_2 stream is then compressed and ready to be transported for geologic sequestration, EOR, or other commercial applications.

1 M NaOH (aq) is a typical solvent concentration because this concentration is limited by the causticization reaction that regenerates the solvent and it is not too far from the practical maximum of 2 M NaOH. The furnace/kiln can be powered renewably or by burning fuel on-site with pure oxygen produces in an on-site air separation unit.

NaOH is economically competitive with other absorbents--e.g., amines--used for DAC processes. DAC processes are energy intensive. Calcination (at the furnace) is the most energy intensive step of this process.

Economic Issues

A crucial issue for CDR methods is their cost, which differs substantially among the different technologies: some of these are not sufficiently developed to perform cost assessments. In 2011 the American Physical Society estimated the costs for direct air capture to be $600/tonne with optimistic assumptions. A 2018 study found this estimate lowered to between $94 and $232 per tonne. The IEA Greenhouse Gas RandD Programme and Ecofys provides an estimate that 3.5 billion tonnes could be removed annually from the atmosphere with BECCS (Bio-Energy with Carbon Capture and Storage) at carbon prices as low as €50 per tonne, while a report from Biorecro and the Global Carbon Capture and Storage Institute estimates costs "below €100" per tonne for large scale BECCS deployment.

Risks, Problems and Criticisms

CDR is slow to act, and requires a long-term political and engineering program to effect. CDR is even slower to take effect on acidified oceans. In a Business as usual concentration pathway, the deep ocean will remain acidified for centuries, and as a consequence many marine species are in danger of extinction.

The Special 1.5°C IPCC report was very clear about CDR: "CDR deployed at scale is unproven and reliance on such technology is a major risk in the ability to limit warming to 1.5°C."

MOBILE EMISSION REDUCTION CREDIT

A mobile emission reduction credit (MERC) is an emission reduction credit generated within the transportation sector. The term "mobile sources" refers to motor vehicles, engines, and equipment that move, or can be moved, from place to place. Mobile sources include vehicles that operate on roads and highways ("on-road" or "highway" vehicles), as well as nonroad vehicles, engines, and equipment. Examples of mobile sources are passenger cars, light trucks, large trucks, buses, motorcycles, earth-moving equipment, nonroad recreational vehicles (such as dirt bikes and snowmobiles), farm and construction equipment, cranes, lawn and garden power tools, marine engines, ships, railroad locomotives, and airplanes. In California, mobile sources account for about 60 percent of all ozone forming emissions and for over 90 percent of all carbon monoxide (CO) emissions from all sources.

Government agencies worldwide have struggled with finding new and innovative approaches to address the growing problem of air pollution and global warming. Experts in the field have recognized the importance of developing solutions to reduce greenhouse gas (GHG) emissions. Most proposed strategies to mitigate global climate change focus on reducing the dominant source of GHG emissions to the atmosphere – combustion of fossil fuels, which releases carbon dioxide. Carbon dioxide emissions represent about 84 percent of total U.S. GHG emissions. In the United States, most carbon dioxide (98 percent) is emitted as a result of the combustion of fossil fuels; consequently, carbon dioxide emissions and energy use are highly correlated.

General Emission Reduction Strategies

The two main approaches that have been developed to address this problem include a command-and-control regulatory system and Emissions credit trading. Three broad types of emissions credit trading programs have emerged: reduction credit, averaging, and cap-and-trade programs. In such programs, a central authority, such as an air pollution control district or a government agency, sets limits or "caps" on certain pollutants. Companies or fleets of vehicles that intend to exceed these limits may buy emission reduction credits (ERCs) from entities that are able to remain below the designated limits. This transfer is usually referred to as a trade.

International Approach to Emission Reduction Credits

Emission trading is contemplated on an international level. The Kyoto Protocol is an agreement made under the United Nations Framework Convention on Climate Change (UNFCCC). The Kyoto Protocol binds ratifying nations to a similar system, with the UNFCCC setting caps for each nation, and utilizes a clean development mechanism (CDM) system. The primary reduction strategy under the Kyoto Protocol is a trading system that essentially makes carbon credits a commodity like oil or gas.

United States Approach to Emission Reduction Credits

The United States (which did not ratify the Kyoto Protocol) has the most experience with domestic emissions trading markets. The Clean Air Act is a federal law that requires the United States Environmental Protection Agency (EPA) to develop and enforce regulations to protect the general public from exposure to airborne contaminants that are known to be hazardous to human health. The Clean Air Act or Clean Air Act amendments of 1990 authorized the use of market-based approaches such as emission trading to assist states in attaining and maintaining air quality for all criteria pollutants. EPA's subsequent interpretive rulings expressly allow owners of new sources to obtain emission credits from other companies that operate facilities located in the same air quality control region. To implement an emissions offset program, many states have developed regulations allowing sources to register their emissions reduction credits as ERCs that can be sold to companies required to offset emissions from new or modified sources. Brokerage companies typically handle sales between companies having surplus ERCs and those wanting to acquire such credits.

All commonly accepted ERCs in the United States must meet each of five criteria before they can be certified by the relevant regulatory authority as an ERC. Namely, the emission reduction must be real, permanent over the period of credit generation, quantifiable, enforceable, and surplus to emission reductions that are already needed to comply with an existing requirement (local, state, or Federal) or air quality plan. These criteria are intended to ensure that the emission reduction is a permanent reduction from the emissions that would otherwise be allowed to offset the permanent increase in emissions from the new or expanding source.

Steps to Create a MERC

The steps involved to create a MERC are as follows:

- Identifying an emissions reduction technology for a pollutant.

- Identifying a mobile source.

- Utilize a Portable Emissions Measurement System to measure emissions of the pollutant and take first measurements of the pollutant from the mobile source.

- Analyze the measurements to develop a baseline emissions amount.

- Apply the emissions reduction technology to the mobile source to provide a modified mobile source.

- Connect the Portable Emissions Measurement System to the modified mobile source and take second measurements of the modified mobile source.

- Analyze the second measurements to develop a modified emissions amount.

- Quantify the mobile emissions reduction produced by the emissions reduction technology.

- Convert the mobile emissions reduction into a tradable commodity.

Monetization of a MERC

The process of converting the mobile emissions reduction into a tradable commodity consists of converting the reduction or a portion of the reduction of emissions into at least one tradable credit, and marketing and monetizing the credit. This is followed by receiving information to identify a customer account, assigning the mobile emissions reduction to the customer account, calculating a MERC from the mobile emissions reduction, and crediting the MERC to the customer account. What follows is the exchanging of the MERC in the customer account for monetary assets this includes the following steps:

- Debiting the MERC from the customer account.

- Receiving information to identify a second customer or purchaser.

- Calculating an emissions amount of the pollutant for the purchaser.

- Assigning a liability value to the emissions amount for the purchaser.

- Accepting payment from the purchaser.

- Using the payment to purchase at least one MERC for the purchaser.

- Crediting the MERC as assets against the liability value assigned to the second customer for the emissions amount, whereby the emissions amount and the liability value in the second customer account is reduced accordingly.

Target Pollutants of Mobile Emission Reduction Credits

At present, the pollutant may be selected from a group consisting of nitrogen oxides (NO_x), carbon monoxides (CO), carbon dioxides (CO_2), hydrocarbons (HC), sulfur oxides (SO_x), particulate matter (PM) and volatile organic compounds (VOCs). The emissions reduction technology may be selected from a group consisting of alternative fuels, vehicle repairs, vehicle replacements, vehicle retrofits and hybrid engines. The mobile source may be selected from a group consisting of passenger cars, light trucks, large trucks, buses, motorcycles, off-road recreational vehicles, farm equipment, construction equipment, lawn and garden equipment, marine engines, aircraft, locomotives and water vessels.

FOSSIL FUEL PHASE-OUT

Fossil fuel phase-out is the gradual reduction of the use of fossil fuels to zero use. Current efforts in fossil fuel phase-out involve replacing fossil fuels with alternative energy sources in sectors such as transport, heating and industry.

Protest at the Legislative Building in Olympia, Washington. Ted Nation an activist for several decades beside protest sign.

The 1968 Farmington coal mine disaster kills 78 in West Virginia, US.

The 2010 Deepwater Horizon oil spill discharges 4.9 million barrels.

Coal

As of 2017 coal supplied over a quarter of the world's primary energy and about 40% of the greenhouse gas emissions from fossil fuels. Although many, but not all, agree that to meet the 2 °C target in the Paris Agreement coal must be phased out worldwide there is much disagreement about how quickly this should be done.

As of 2018, 30 countries and many sub-national governments and businesses had become members of the Powering Past Coal Alliance, each making a declaration to advance the transition away from unabated coal power generation. As of 2019, however, the countries which use the most coal have not joined, and some countries continue to build and finance new coal-fired power stations.

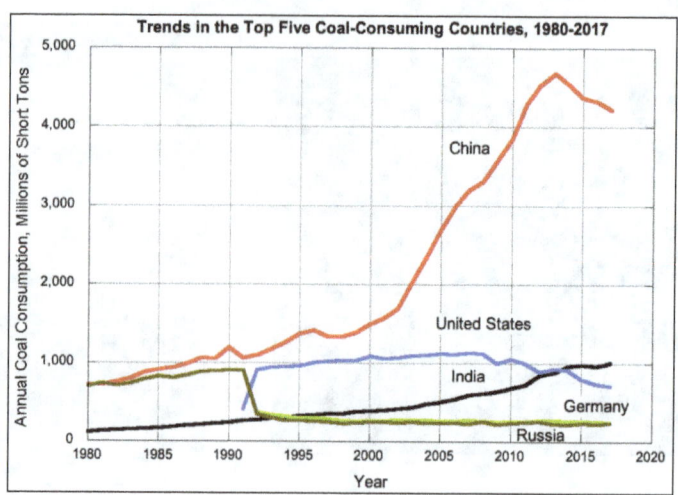

Top 5 coal consuming countries to 2017, US EIA data.

In 2019 the UN Secretary General said that countries should stop building new coal power plants from 2020 or face 'total disaster'.

Coal-fired power plants provided 30% of consumed electricity in the United States in 2016. This is the Castle Gate Plant near Helper, Utah.

Oil

Oil is refined into fuel oil, diesel and gasoline. The refined products are primarily for transportation by conventional cars, trucks, trains, planes and ships. Popular alternatives are human-powered transport, public transport, electric vehicles, and biofuels.

Natural Gas

Natural gas is widely used to generate electricity and has an emission intensity of about 500g/kWh. Heating is also a major source of carbon dioxide emissions. Leaks are also a large source of atmospheric methane.

In some countries natural gas is being used as a temporary "bridge fuel" to replace coal, in turn to be replaced by renewable sources or a hydrogen economy. However this "bridge fuel" may significantly extend the use of fossil fuel or strand assets, such as gas-fired power plants built in the 2020s, as the average plant life is 35 years. Although natural gas assets are likely to be stranded later than oil and coal assets, perhaps not until 2050, some investors are concerned by reputational risk.

Natural gas phase-out is progressing in some regions, for example with increasing use of hydrogen by the European Network of Transmission System Operators for Gas (ENTSOG) and changes to building regulations to reduce the use of gas heating.

The reasons for phasing-out fossil fuels are: the health risks of air pollution, mitigation of global warming, and the falling cost of renewable energy.

Health

Most of the millions of premature deaths from air pollution are due to fossil fuels. Pollution may be indoors e.g. from heating and cooking, or outdoors from vehicle exhaust. One estimate is that the proportion is 65% and the number 3.5 million each year. According to Professor Sir Andy Haines at the London School of Hygiene and Tropical Medicine the health benefits of phasing out fossil fuels measured in money (estimated by economists using the value of life for each country) are substantially more than the cost of achieving the 2 degree C goal of the Paris Agreement.

Global Warming Mitigation

More recently, Hansen has stated that continued opposition to nuclear power threatens humanity's ability to avoid dangerous climate change. The letter, co-authored with other climate change experts declared "If we stay on the current path," he said, "those are the consequences we'll be leaving to our children. The best candidate to avoid that is nuclear power. It's ready now. We need to take advantage of it." and "Continued opposition to nuclear power threatens humanity's ability to avoid dangerous climate change."

Also in 2008, Pushker Kharecha and James Hansen published a peer-reviewed scientific study analyzing the effect of a coal phase-out on atmospheric carbon dioxide (CO_2) levels. Their baseline mitigation scenario was a phaseout of global coal emissions by 2050. The authors describe the scenario as follows:

> The second scenario, labeled Coal Phase-out, is meant to approximate a situation in which developed countries freeze their CO_2 emissions from coal by 2012 and a decade later developing countries similarly halt increases in coal emissions. Between 2025 and 2050 it is assumed that both developed and developing countries will linearly phase out emissions of CO_2 from coal usage. Thus in Coal Phase-out we have global CO_2 emissions from coal increasing 2% per year until 2012, 1% per year growth of coal emissions between 2013 and 2022, flat

coal emissions for 2023–2025, and finally a linear decrease to zero CO_2 emissions from coal in 2050. These rates refer to emissions to the atmosphere and do not constrain consumption of coal, provided the CO_2 is captured and sequestered. Oil and gas emissions are assumed to be the same as in the BAU (business as usual) scenario.

Kharecha and Hansen also consider three other mitigation scenarios, all with the same coal phase-out schedule but each making different assumptions about the size of oil and gas reserves and the speed at which they are depleted. Under the Business as Usual scenario, atmospheric CO_2 peaks at 563 parts per million (ppm) in the year 2100. Under the four coal phase-out scenarios, atmospheric CO_2 peaks at 422-446 ppm between 2045 and 2060 and declines thereafter. The key implications of the study are as follows: a phase-out of coal emissions is the most important remedy for mitigating human-induced global warming; actions should be taken toward limiting or stretching out the use of conventional oil and gas; and strict emissions-based constraints are needed for future use of unconventional fossil fuels such as methane hydrates and tar sands.

Others

In the Greenpeace and EREC's Energy Revolution scenario, the world would eliminate all fossil fuel use by 2090.

In December 2015, Greenpeace and Climate Action Network Europe released a report highlighting the need for an active phase-out of coal-fired generation across Europe. Their analysis derived from a database of 280 coal plants and included emissions data from official EU registries.

A September 2016 report by Oil Change International, concludes that the carbon emissions embedded in the coal, oil, and gas in currently working mines and fields, assuming that these run to the end of their working lifetimes, will take the world to just beyond the 2 °C limit contained in the 2015 Paris Agreement and even further from the 1.5 °C goal. The report observes that "one of the most powerful climate policy levers is also the simplest: stop digging for more fossil fuels".

In October 2016, the Overseas Development Institute (ODI) and 11 other NGOs released a report on the impact of building new coal-fired power plants in countries where a significant proportion of the population lacks access to electricity. The report concludes that, on the whole, building coal-fired power plants does little to help the poor and may make them poorer. Moreover, wind and solar generation are beginning to challenge coal on cost.

A 2018 study in Nature Energy, suggests that 10 countries in Europe could completely phase out coal-fired electricity generation with their current infrastructure, whilst the United States and Russia could phase out at least 30%.

Challenges of Fossil Fuel Phase-out

The phase-out of fossil fuels involves many challenges, and one of them is the reliance that currently the world has on them. In 2014, fossil fuels provided 81.1% of the primary energy consumption of the world, with approximately 11,109 megatonnes of oil equivalent. This number is composed by 4,287 Mtoe of oil consumption; 3,918 Mtoe of coal consumption, and 2,904 Mtoe of natural gas consumption.

Fossil fuel phase-out can lead to an increment in electricity prices, because of the new investments needed to replace their share in the electricity mix with alternative energy sources. Another cause to increasing electricity price comes from the need to import the electricity that can't be generated nationally.

Another impact of a phase-out of fossil fuels is in the employment. In the case of employments in the fossil fuel industry, a phase-out is logically undesired, therefore, people in the industry will usually oppose any measures that put their industries under scrutiny. Endre Tvinnereim and Elisabeth Ivarsflaten studied the relationship between employment in the fossil fuel industry with the support to climate change policies. They proposed that one opportunity for displaced drilling employments in the fossil fuel industry could be in the geothermal energy industry. This was suggested as a result of their conclusion: people and companies in the fossil fuel industry will likely oppose measures that endanger their employments, unless they have other stronger alternatives. This can be extrapolated to political interests, that can push against the phase-out of fossil fuels initiative. One example is how the vote of US Congress members is related to the preeminence of fossil fuel industries in their respective states.

Phase-out of Fossil Fuel Vehicles

Many countries and cities have introduced bans on the sales of new internal combustion engine vehicles, requiring all new cars to be electric vehicles or otherwise powered by clean, non-emitting sources. Such bans include the United Kingdom by 2040 and Norway by 2025. Many transit authorities are working to purchase only electric buses while also restricting use of ICE vehicles in the city center to limit air pollution. Many US states have a zero-emissions vehicle mandate, incrementally requiring a certain percent of cars sold to be electric.

Alternative Sources of Energy

Alternative energy refers to any source of energy that can substitute the role of fossil fuels. Renewable energy, or energy that is harnessed from renewable sources, is an alternative energy. However, alternative energy can refer to non renewable sources as well, like nuclear energy. Between the alternative sources of energy are: solar energy, hydroelectricity, marine energy, wind energy, geothermal energy, biofuels, ethanol and Hydrogen.

Energy efficiency is complementary to the use of alternative energy sources, when phasing-out fossil fuels.

Renewable Energy

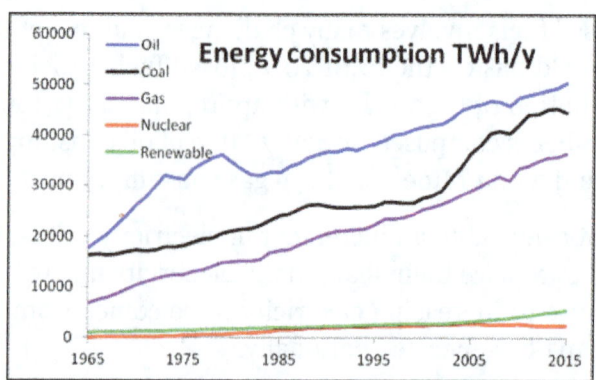

The worldwide growth of renewable energy is shown by the green line.

Renewable energy is energy that comes from resources which are naturally replenished such as sunlight, wind, rain, tides, waves, and geothermal heat. As of 2014, 19% of global final energy consumption comes from renewable resources, with 9% of all energy from traditional biomass, mainly used for heating, 1% from biofuels, 4% from hydroelectricity and 4% from biomass, geothermal or solar heat. Popular renewables (wind, solar, geothermal and biomass for power) accounted for another 1.4% and are growing rapidly. While renewable energy supplies are growing and have displaced coal in some regions, the amount of coal burned in 2021, is expected to be the same as it was in 2014.

Hydroelectricity

Chief Joseph Dam near Bridgeport, Washington, US, is a major
run-of-the-river station without a sizeable reservoir.

In 2015, hydroelectric energy generated 16.6% of the world's total electricity and 70% of all renewable electricity. In Europe and North America environmental concerns around land flooded by large reservoirs ended 30 years of dam construction in the 1990s. Since then large dams and reservoirs continue to be built in countries like China, Brazil and India. Run-of-the-river hydroelectricity and small hydro have become popular alternatives to conventional dams that may create reservoirs in environmentally sensitive areas.

Wind Power

A wind farm is a group of wind turbines in the same location used to produce electric power. A large wind farm may consist of several hundred individual wind turbines, and cover an extended area of hundreds of square miles, but the land between the turbines may be used for agricultural or other purposes. A wind farm may also be located offshore.

Lillgrund Wind Farm in Sweden.

Wind power has grown dramatically since 2005 and by 2015 supplied almost 1% of global energy consumption.

First wind farm consisting of 7,5 MW Enercon E-126 turbines, Estinnes,
Belgium, 20 July 2010, two months before completion; note the 2-part blades.

Many of the largest operational onshore wind farms are located in the United States and China. The Gansu Wind Farm in China has over 5,000 MW installed with a goal of 20,000 MW by 2020. China has several other "wind power bases" of similar size. The Alta Wind Energy Center in California, United States is the largest onshore wind farm outside of China, with a capacity of 1020 MW of power. As of February 2012, the Walney Wind Farm in the United Kingdom is the largest offshore wind farm in the world at 367 MW, followed by Thanet Offshore Wind Project (300 MW), also in the United Kingdom. As of February 2012, the Fântânele-Cogealac Wind Farm in Romania is the largest onshore wind farm in Europe at 600 MW.

There are many large wind farms under construction and these include Sinus Holding Wind Farm (700 MW), Anholt Offshore Wind Farm (400 MW), BARD Offshore 1 (400 MW), Clyde Wind Farm (350 MW), Greater Gabbard wind farm (500 MW), Lincs Wind Farm (270 MW), London Array (1000 MW), Lower Snake River Wind Project (343 MW), Macarthur Wind Farm (420 MW), Shepherds Flat Wind Farm (845 MW), and Sheringham Shoal (317 MW).

Wind power in Denmark produced the equivalent of 42.1% of total electricity consumption in 2015, however, use of wind for heating is minor.

Solar

In 2017, solar power provided 1.7% of total worldwide electricity production, growing at 35% per annum. By 2020 the solar contribution to global final energy consumption is expected to exceed 1%.

Solar Photovoltaics

Solar photovoltaic cells convert sunlight into electricity and many solar photovoltaic power stations have been built. The size of these stations has increased progressively over the last decade with frequent new capacity records. Many of these plants are integrated with agriculture and some use innovative tracking systems that follow the sun's daily path across the sky to generate more electricity than conventional fixed-mounted systems. Solar power plants have no fuel costs or emissions during operation.

The 71.8 MW Lieberose Photovoltaic Park in Germany.

Concentrated Solar Power

Concentrating Solar Power (CSP) systems use lenses or mirrors and tracking systems to focus a large area of sunlight into a small beam. The concentrated heat is then used as a heat source for a conventional power plant. A wide range of concentrating technologies exists; the most developed are the parabolic trough, the concentrating linear fresnel reflector, the Stirling dish and the solar power tower. Various techniques are used to track the Sun and focus light. In all of these systems a working fluid is heated by the concentrated sunlight, and is then used for power generation or energy storage.

The 150 MW Andasol solar power station is a commercial parabolic trough solar thermal power plant, located in Spain. The Andasol plant uses tanks of molten salt to store solar energy so that it can continue generating electricity even when the sun isn't shining.

Biofuels

Biofuels, in the form of liquid fuels derived from plant materials, are entering the market. However, many of the biofuels that are currently being supplied have been criticised for their adverse impacts on the natural environment, food security, and land use.

Biomass

Biomass is biological material from living, or recently living organisms, most often referring to plants or plant-derived materials. As a renewable energy source, biomass can either be used directly, or indirectly – once or converted into another type of energy product such as biofuel. Biomass can be converted to energy in three ways: thermal conversion, chemical conversion, and biochemical conversion.

Using biomass as a fuel produces air pollution in the form of carbon monoxide, carbon dioxide, NO_x (nitrogen oxides), VOCs (volatile organic compounds), particulates and other pollutants at levels above those from traditional fuel sources such as coal or natural gas in some cases (such as with indoor heating and cooking). Utilization of wood biomass as a fuel can also produce fewer particulate and other pollutants than open burning as seen in wildfires or direct heat applications. Black carbon – a pollutant created by combustion of fossil fuels, biofuels, and biomass – is possibly the second largest contributor to global warming. In 2009 a Swedish study of the giant brown haze that periodically covers large areas in South Asia determined that it had been principally produced by biomass burning, and to a lesser extent by fossil fuel burning. Denmark has increased the use of biomass and garbage, and decreased the use of coal.

Nuclear Energy

The 2014 Intergovernmental Panel on Climate Change report identifies nuclear energy as one of the technologies that can provide electricity with less than 5% of the lifecycle greenhouse gas emissions of coal power. There are more than 60 nuclear reactors shown as under construction in the list of Nuclear power by country with China leading at 23. Globally, more nuclear power reactors have closed than opened in recent years but overall capacity has increased. China has stated its plans to double nuclear generation by 2030. India also plans to greatly increase its nuclear power.

Several countries have enacted laws to cease construction on new nuclear power stations. Several European countries have debated nuclear phase-outs and others have completely shut down some reactors. Three nuclear accidents have influenced the slowdown of nuclear power: the 1979 Three Mile Island accident in the United States, the 1986 Chernobyl disaster in the USSR, and the 2011 Fukushima nuclear disaster in Japan. Following the March 2011 Fukushima nuclear disaster, Germany has permanently shut down eight of its 17 reactors and pledged to close the rest by the end of 2022. Italy voted overwhelmingly to keep their country non-nuclear. Switzerland and Spain have banned the construction of new reactors. Japan's prime minister has called for a dramatic reduction in Japan's reliance on nuclear power. Taiwan's president did the same. Shinzō Abe, prime minister of Japan since December 2012, announced a plan to restart some of the 54 Japanese nuclear power plants and to continue some nuclear reactors under construction.

As of 2016, countries such as Australia, Austria, Denmark, Greece, Malaysia, New Zealand, and Norway have no nuclear power stations and remain opposed to nuclear power. Germany, Italy, Spain and Switzerland are phasing-out their nuclear power.

Energy Efficiency

Moving away from fossil fuels will require changes not only in the way energy is supplied, but in the way it is used, and reducing the amount of energy required to deliver various goods or services is essential. Opportunities for improvement on the demand side of the energy equation are as rich and diverse as those on the supply side, and often offer significant economic benefits.

A sustainable energy economy requires commitments to both renewables and efficiency. Renewable energy and energy efficiency are said to be the "twin pillars" of sustainable energy policy. The American Council for an Energy-Efficient Economy has explained that both resources must be developed in order to stabilize and reduce carbon dioxide emissions:

> Efficiency is essential to slowing the energy demand growth so that rising clean energy supplies can make deep cuts in fossil fuel use. If energy use grows too

fast, renewable energy development will chase a receding target. Likewise, unless clean energy supplies come online rapidly, slowing demand growth will only begin to reduce total emissions; reducing the carbon content of energy sources is also needed.

The IEA has stated that renewable energy and energy efficiency policies are complementary tools for the development of a sustainable energy future, and should be developed together instead of being developed in isolation.

POLICIES, INSTRUMENTS AND CO-OPERATIVE AGREEMENTS

A variety of policies, measures, instruments and approaches are available to national governments to limit the emission of GHGs; these include regulations and standards, taxes and charges, tradable permits, voluntary agreements (VAs), informational instruments, subsidies and incentives, research and development and trade and development assistance. Depending on the legal framework within which each individual country must operate, these may be implemented at the national level, sub-national level or through bi-lateral or multi-lateral arrangements, and they may be either legally binding or voluntary and either fixed or changeable (dynamic).

Criteria for Policy Choice

- Four principal criteria for evaluating environmental policy instruments are:

- Environmental effectiveness: The extent to which a policy meets its intended environmental objective or realizes positive environmental outcomes.

- Cost-effectiveness: The extent to which the policy can achieve its objectives at a minimum cost to society.

- Distributional considerations: The incidence or distributional consequences of a policy, which includes dimensions such as fairness and equity, although there are others.

- Institutional feasibility: The extent to which a policy instrument is likely to be viewed as legitimate, gain acceptance, adopted and implemented.

Environmental Effectiveness

The main goal of environmental policy instruments and international agreements is to reduce the negative impact of human action on the environment. Policies that achieve specific environmental quality goals better than alternative policies can be said to have a higher degree of environmental effectiveness. It should be noted that although climate protection is the ostensible environmental goal for any climate policy, there may be ancillary environmental benefits.

The environmental effectiveness of any policy is contingent on its design, implementation, participation, stringency and compliance. For example, a policy that seeks to fully address the climate problem while dealing with only some of the GHGs or some of the sectors will be relatively less effective than one that aims at addressing all gases and all sectors.

The environmental effectiveness of an instrument can only be determined by estimating how well it is likely to perform. Harrington et al. distinguish between estimating how effective an environmental instrument will be ex ante and evaluating its performance ex post. These researchers were able to find or recreate ex ante estimates of expected emissions reductions in a series of U.S. and European case studies. Their comparison of the ex ante and ex post observations suggests a reasonable degree of accuracy in the estimates, with those cases in which emissions reductions were greater than expected involving incentive-based instruments, while the cases in which reductions fell short of expectations involved regulatory approaches.

There are situations in which standards are proven to be effective. Regulators may be unduly pessimistic about the environmental performance of incentive-based instruments or unduly optimistic about the performance of regulatory approaches, or perhaps both. Recent evidence suggests that market-based approaches can provide equal if not superior environmental quality improvements over regulatory approaches. As we discuss below, however, institutional constraints may alter the relative efficacy of market- and standards-based instruments.

Cost-effectiveness

The cost-effectiveness of a policy is a key decision parameter in a world with scarce resources. Given a particular environmental quality goal, the most cost-effective policy is the one which achieves the desired goal at the least cost. There are many components of cost, and these include both the direct costs of administering and implementing the policy as well as indirect costs, such as how the policy drives cost-reducing technological change.

Cost-effectiveness is distinct from general economic efficiency. Whereas cost-effectiveness takes an environmental goal as given, efficiency involves the process of selecting a specific goal according to economic criteria. Consequently, the choice of a particular environmental goal will likely have dramatic impacts on the overall cost of a policy, even if that policy is implemented using the most cost-effective instrument.

Policies are likely to vary considerably in terms of costeffectiveness, and any estimation of the costs involved can be challenging. While cost-effectiveness estimates traditionally include the direct expenditures incurred as a result of implementing any specific policy, the policy may also impose indirect social costs, which are more difficult to measure. Moreover, costs for which data are limited are often ignored. Harrington et al. provide a summary of commonly excluded costs as well as examples of efforts to estimate these.

Cost-effectiveness can be enhanced with low transaction costs for compliance. This implies limiting the creation of new institutions and keeping implementation procedures as simple as possible while still ensuring system integrity. Studies reported in the literature can be divided into two categories in terms of the economic impacts of the timing of reductions. While some researchers argue that reductions should be postponed until low-cost technologies are available, others argue that necessary decisions have to be made today to avoid a 'lock-in' to an emission intensive pathway that would be expensive to leave at a later time point.

A common concern is that ex ante cost estimates may not reflect the actual costs of a policy when it is assessed from an ex post perspective. Harrington et al. show that the discrepancy between the actual and estimated total costs of 28 environmental regulations in the USA is relatively low and, if anything, that ex ante estimates tend to overstate total costs. While these authors do not systematically evaluate specific environmental instruments, they do find that estimates for market-based instruments tend to overstate unit costs, while unit-costs estimates for other instruments are neither under- nor overestimates.

Distributional Considerations

Policies rarely apportion environmental benefits and costs evenly across stakeholders. Even if a policy meets an environmental goal at least cost, it may face political opposition if it disproportionately impacts – or benefits – certain groups within a society, across societies or across generations. From an economic perspective, a policy is considered to be beneficial if it improves social welfare overall. However, this criterion does not require that the implementation of that policy actually improves the specific situation of any one individual. Consequently, as Keohane et al. argue, distributional considerations may be more important than aggregate cost effectiveness when policymakers evaluate an instrument.

The distributional considerations of climate change policies relate largely to equity. Equity can be defined in a number of ways within the climate context. Equity and fairness may be perceived differently by different people, depending on the cultural background of the observer. For example, Ringius et al. view responsibility, capacity and need as the basic principles of fairness that seem to be sufficiently widely recognized to serve as a normative basis for a climate policy regime. These three principles have been used in the evaluation of potential international climate agreements.

A regulation that is perceived as being unfair or for which the incidence is unbalanced may have a difficult time making it through the political process. However, distributional considerations are fundamentally subjective, and the most equitable policy may not be the most politically popular one. For example, a policy that focuses the regulatory burden on a low-income subpopulation or country but directs the benefits to a wealthy interest group may sail with ease through the political process. While highly inequitable in costs and benefits, such an instrument is occasionally attractive to politicians. Bulkeley describes the different interests in the Australian climate policy debate and suggests that industrial emitters managed to steer the country away from ambitious reduction target – and toward an emissions increase – at the third Conference of the Parties in Kyoto.

Due to the fact that there is little consensus as to what constitutes optimal distribution, it can be difficult to compare – let alone rank – environmental policies based on distributional criteria. One exception is provided by Asheim et al., who construct an axiom of equity which, they argue, can be used to evaluate sustainability. However, while sustainability may be important when evaluating environmental policies, it only captures the inter-generational dimension of distribution and is imperfectly related to political acceptability.

Institutional Feasibility

Institutional realities inevitably constrain environmental policy decisions. Environmental policies that are well adapted to existing institutional constraints have a high degree of institutional feasibility. Economists traditionally evaluate instruments for environmental policy under ideal theoretical conditions; however, those conditions are rarely met in practice, and instrument design and implementation must take political realities into account. In reality, policy choices must be both acceptable to a wide range of stakeholders and supported by institutions, notably the legal system. Other important considerations include human capital and infrastructure as well as the dominant culture and traditions. The decision-making style of each nation is therefore a function of its unique political heritage.

Certain policies may also be popular due to institutional familiarity. Although market-based instruments are becoming more common, they have often met with resistance from environmental groups. Market-based instruments continue to face strong political opposition, even in the developed world, as demonstrated by environmental taxes in the USA or Europe. Regulatory policies that are outside of the norm of society will always be more difficult to put into effect (e.g. speed limits in Germany, or private sector participation in water services in Bolivia).

Another important dimension of institutional feasibility deals with implementing policies once they have been designed and adopted. Even if a policy receives political support, it may be difficult to implement under certain bureaucratic structures.

National Policy Instruments, their Implementation and Interactions

The policy-making process of almost all governments consists of complex choices involving many stakeholders, including the potential regulated industry, suppliers, producers of complementary products, labour organizations, consumer groups and environmental organizations. The choice and design of virtually any instrument has the potential to benefit some of these stakeholders and to harm others. For example, permits allocated free to existing firms represent a valuable asset transferred from the government to industry, while auctioned permits and taxes generally impose heavier burdens on polluters. As a result, it is likely that a candidate instrument will likely face both support and opposition from the stakeholders. Voluntary measures are often favoured by industry because of their flexibility and potentially lower costs, but these are often opposed by environment groups because of their lack of accountability and enforcement. In practice, policies may be complementary or opposing; moreover, the political calculus used to choose a particular instrument differs for each government.

In formulating a domestic climate policy programme, a combination of policy instruments may work better in practice than reliance on a single instrument. Furthermore, an instrument that works well in one country may not work well in another country with different social norms and institutions. When instruments are to be compared, it is important that the different levels of stringency be taken into consideration and adjusted, for all of the instruments described herein may be set at different levels of stringency. Regulations will also undoubtedly need to be adjusted over time. All instruments must be supplemented with a workable system of monitoring and enforcement. Furthermore, instruments may interact with existing institutions and regulations in other sectors of society.

References

- "Goal 7—Ensure Access to Affordable, Reliable, Sustainable and Modern Energy for All". UN Chronicle. 8 April 2015. Retrieved 27 September 2019

- Detection-of-grenhouse-gases-using-the-photoacoustic-spectroscopy, greenhouse-gases-emission-measurement-and-management, books: intechopen.com, Retrieved 15 April, 2020

- Constanze Werner et al. (2018): Biogeochemical potential of biomass pyrolysis systems for limiting global warming to 1.5° C. Environmental Research Letters, 13(4), 044036. doi:10.1088/1748-9326/aabb0e

- Huesemann, Michael H., and Joyce A. Huesemann (2011). Technofix: Why Technology Won't Save Us or the Environment, Chapter 5, "In Search of Solutions: Efficiency Improvements", New Society Publishers, ISBN 978-0-86571-704-6

- "About Transportation & Climate Change: Transportation's Role in Climate Change: Overview - DOT Transportation and Climate Change Clearinghouse". climate.dot.gov. Retrieved 2015-11-15

- Pearce, Joshua M.; Johnson, Sara J.; Grant, Gabriel B. (2007). "3D-Mapping Optimization of Embodied Energy of Transportation". Resources, Conservation and Recycling. 51 (2): 435–453. doi:10.1016/j.resconrec.2006.10.010. Retrieved 1 March 2018

Carbon neutrality refers to the process of nullifying carbon emission with carbon removal to achieve net zero carbon dioxide emissions. It includes concepts of carbon accounting, carbon offset, carbon sequestration, carbon capture and storage, carbon footprint, etc. This chapter covers all the related concepts of carbon neutrality for a thorough understanding of the subject.

Carbon neutrality, or having a net zero carbon footprint, refers to achieving net zero carbon emissions by balancing a measured amount of carbon released with an equivalent amount sequestered or offset, or buying enough carbon credits to make up the difference. It is used in the context of carbon dioxide releasing processes, associated with transportation, energy production and industrial processes.

The carbon neutral concept may be extended to include other greenhouse gases (GHG) measured in terms of their carbon dioxide equivalence—the impact a GHG has on the atmosphere expressed in the equivalent amount of CO_2. The term climate neutral is used to reflect the fact that it is not just carbon dioxide (CO_2), that is driving climate change, even if it is the most abundant, but also encompasses other greenhouse gases regulated by the Kyoto Protocol, namely: methane (CH_4), nitrous oxide (N_2O), hydrofluorocarbons (HFC), perfluorocarbons (PFC), and sulphur hexafluoride (SF_6).

Best practice for organizations and individuals seeking carbon neutral status entails reducing and avoiding carbon emissions first so that only unavoidable emissions are offset.

The term has two common uses:

It can refer to the practice of balancing carbon dioxide released into the atmosphere from burning fossil fuels, with renewable energy that creates a similar amount of useful energy, so that the carbon emissions are compensated, or alternatively using only renewable energies that don't produce any carbon dioxide.

It is also used to describe the practice, criticized by some, of carbon offsetting, by paying others to remove or sequester 100% of the carbon dioxide emitted from the atmosphere– for example by planting trees – or by funding 'carbon projects' that should lead to the prevention of future greenhouse gas emissions, or by buying carbon credits to remove them through carbon trading. These practices are often used in parallel, together with energy conservation measures to minimize energy use.

Achieving carbon neutrality entails the completion of the following three separate stages:

Calculating Emissions

This stage requires the determination of what emissions will be calculated, including setting a clear boundary for emissions covered (in terms of the gases included, the organisational context and the sources of emissions). Once the boundary has been set, emissions can be calculated by collecting activity data (for example, the amount of electricity and gas consumed) and applying the appropriate emissions factors.

Reducing Emissions

This stage involves assessing what internal emissions reductions can be made through e.g. energy efficiency measures. These will usually be carried out because they are cost effective over time, helping to save money at the same time as reducing emissions. Reductions can be based on absolute emission reductions or emission reductions relative to a common business metric or unit of output. Those seeking to become carbon neutral should decide how to reduce emissions, how to calculate reductions and how to communicate this.

Offsetting Residual Emissions

This third stage requires the acquisition of carbon credits to offset any residual emissions after calculating emissions and achieving internal reductions. The precise amount of offsets required needs to be calculated, with enough credits bought to reduce emissions to net zero. When offsetting, consideration should be given to the type of offsets bought to be sure they are good quality and represent a real (tonne for tonne) emissions reduction.

Communicating Carbon Neutrality

In making a carbon neutral statement, the business or individual should provide clear information on the emissions measured, the reductions made and the offsets purchased. The same applies for products. Taking this approach will allow consumers and interested parties to view each claim on its own merits.

A claim of carbon neutrality should always be linked to a particular and specified period of time because doing so will ensure the claim is understandable and transparent. It also helps ensure that carbon neutral efforts are regularly reviewed and updated.

It is recommended that carbon neutral statements should be reviewed yearly. This period of review aligns with existing reporting systems such as annual and corporate reporting.

It is also recommended that organisations, groups and individuals should be clear about the emissions measured. An organisation that measures emissions from only limited sources must be clear that this is the case. For example, if a carbon neutral taxi business has calculated emissions from owned transport but not from energy use in office premises, it should be clear that the term carbon neutral only applies to its vehicle fleet.

Carbon offsets representing genuine, additional emissions reductions made elsewhere have a role in helping groups and individuals to become carbon neutral by addressing residual emissions and ensuring that overall net emissions are equal to zero. Offsetting should, however, be viewed within the hierarchy of action to tackle climate change referred to above. In other words, always seek to reduce emissions to minimise the emissions that need to be offset. A carbon neutral claim consisting only of calculating emissions and offsetting should not be made.

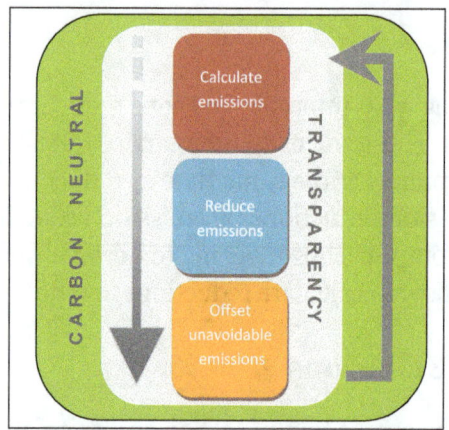

Graphical representation of carbon neutrality.

It is recommended that users of the term verify their carbon neutral statements. Verification would reinforce efforts to make transparent claims. Such verification could be undertaken to confirm emissions reductions achieved or the quality of carbon offsets bought as well as the carbon neutral statement in its entirety. There are many organisations that provide a verification service.

Calculating Emissions

The foundation to any carbon neutrality claim is calculating emissions in an accurate, consistent and transparent way.

Categorising your Emissions

The most common and internationally accepted approach to categorising emissions is through the Greenhouse Gas Protocol.

The GHG Protocol Groups Emissions into Three Different 'Scopes'

Scope 1 (Direct emissions): Activities that are owned or controlled that release emissions straight into the atmosphere. They are direct emissions. Examples of scope 1 emissions include emissions from combustion in owned or controlled boilers, furnaces, vehicles owned or controlled; emissions from chemical production in owned or controlled process equipment.

Scope 2 (Energy indirect): Emissions being released into the atmosphere associated with consumption of purchased electricity, heat, steam and cooling. These are indirect emissions that are a consequence of activities which occur at sources not owned or controlled.

Scope 3 (Other indirect): The final category is all other activities that release emissions into the atmosphere as a consequence of actions taken, which occur at sources that are not owned or controlled and which are not classed as scope 2 emissions, i.e. they do not result from the purchase of electricity, heat, steam and cooling. Examples of scope 3 emissions are business travel by means which are not owned or controlled, waste disposal, and use of sold products or services.

Setting Boundaries

When looking to achieve carbon neutrality, you should first set the boundary for the emissions you are going to address and to which any carbon neutrality declaration you make will relate.

The scope of the emissions included in the footprint is up to you as an individual, community or organisation. The boundary of any carbon neutral declaration can be set at the most appropriate point for you. However, 'to avoid claims' being misinterpreted, you should clearly communicate information on what emissions have been included.

Minimum, emissions within both scopes 1 and 2 should be included in any calculation of emissions. In addition, it is recommended that organisations include their significant scope 3 emissions. Calculating scope 3 emissions will mean developing a more complete understanding of the total impact on climate change. However it is acknowledged that it can be difficult to measure and calculate scope 3 emissions. The following criteria have been set out in the Government's guidance for organisations to help identify those scope 3 emissions that are significant:

- Scale: What are the largest indirect emissions-causing activities with which the organisation concerned is connected?

- Importance to business/activity: Are there any sources of GHG emissions that are particularly important to the business/activity in question or that increase climate change risks (e.g. electricity consumption in the case of consumer use of energy using products or emissions from vehicle use for motor manufacturers)?

- Importance to stakeholders: Which emission causing activities do interested parties (e.g. customers, suppliers, investors) expect to see reported?

- Potential for reductions: Where is there potential to influence or reduce emissions from indirect emission activities?

Once the boundaries have been set for a given period, you can calculate total emissions from the relevant sources. The most common approach used is to apply documented emissions factors to known activity data from the organisation. Activity data is information used to calculate GHG emissions from combustion and other processes e.g. the litres of fuel consumed by a car or electricity use in kilowatt hours. Most activity data is easy to obtain, accurate and can be found on bills, invoices and receipts. The period for which data is collected can vary to suit any reporting needs. However we recommend that the period for reporting or reviewing carbon neutrality should be 12 months. Once gathered, the activity data is converted into CO_2e for business and community claims by multiplying the activity data by the relevant emissions factors.

Individuals and families who use the Act on CO_2 calculator to measure their household or individual emissions will have the final measurement presented to them in tonnes of CO_2. In such instances, no further calculations are required.

The boundary set for one carbon neutral period may change for a second carbon neutral period. This can be through choice (perhaps to include the calculation of more emissions), or because of factors such as business expansion. If the scope of emissions being addressed through the carbon neutrality claim changes over time then this should be communicated clearly.

Reducing Emissions

Reducing emissions internally is a vital part of achieving carbon neutrality. Reductions in emissions e.g. through energy efficiency measures or cutting back travel can deliver cost savings.

When using the term carbon neutral you may wish to set an internal reduction target to be achieved over a given time-frame. This guidance does not include specific recommendations on the nature of the reduction target you should set, nor the period in which you should deliver it. However, it is important to stress that reducing emissions internally is a continuous process, not least because innovation is likely to increase the ways in which you will be able to make internal emissions reductions over time. Recognising the need to take internal action as much as possible if UK targets are to be met, and the fact that there is a wide range of cost-effective reduction measures available, it is recommended that any internal reductions or reduction plans be as ambitious as possible.

Internal emissions reduction measures include:

*   Carrying out/completion of projects such as energy efficiency measures, through e.g. installation of on-site renewable, behaviour change programmes and supplier engagement strategies.

*   Generation and consumption of electricity from renewable sources backed by Renewable Energy Guarantees of Origin (REGOs) certificates.

*   The purchase of green tariffs which comply with OFGEM's Independent Certification Scheme.

When reporting a reduction in emissions, it is possible to choose whether the reduction will be reported as an absolute reduction or an intensity based reduction. An example of the former would be a 5% reduction from one year to the next in the total emissions from gas use. An example of the latter would be a 5% reduction from one year to the next in emissions per cubic metre of office space. If you choose an intensity based reduction it should be relevant to your situation. For example, a suitable intensity measurement for reductions in emissions from office premises would be per metre squared. The Government's guidance on measuring and reporting your greenhouse gas emissions includes further examples of possible measures.

CARBON ACCOUNTING

Carbon accounting refers generally to processes undertaken to "measure" amounts of carbon dioxide equivalents emitted by an entity. It is used by states, corporations and individuals to create the carbon credit commodity traded on carbon markets (or to establish the demand for carbon credits). Correspondingly, examples for products based upon forms of carbon accounting can be found in national inventories, corporate environmental reports or carbon footprint calculators. Likening sustainability measurement, as an instance of ecological modernisation discourses and policy, carbon accounting is hoped to provide a factual ground for carbon-related decision-making. However, social scientific studies of accounting challenge this hope, pointing to the socially constructed character of carbon conversion factors or of the accountants' work practice which cannot implement abstract accounting schemes into reality.

While natural sciences claim to know and measure carbon, for organisations it is usually easier to employ forms of carbon accounting to represent carbon. The trustworthiness of accounts of carbon emissions can easily be contested. Thus, how well carbon accounting represents carbon is difficult to exactly know. Science and Technology Studies scholar Donna Haraway's pluralised concept of knowledge, i.e. knowledges, can well be used to understand better the status of knowledge produced by carbon accounting: carbon accounting produced a version of understanding of carbon emissions. Other carbon accountants would produce other results.

Carbon Accounting in Corporations

Carbon accounting can be used as part of sustainability accounting by for-profit and non-profit organisations. A corporate or organisational "carbon" or greenhouse gas (GHG) emissions assessment promises to quantify the greenhouse gases produced directly and indirectly from a business or organisation's activities within a set of boundaries. Also known as a carbon footprint, it is a business tool that constructs information that may (or may not) be useful for understanding and managing climate change impacts.

The drivers for corporate carbon accounting include mandatory GHG reporting in directors' reports, investment due diligence, shareholder and stakeholder communication, staff engagement, green messaging, and tender requirements for business and government contracts. Accounting for greenhouse gas emissions is increasingly framed as a standard requirement for business. As of June 2011, 60% of UK FTSE 100 companies had published environmental targets, with 53% of these 240+ targets relating to carbon, greenhouse gas emissions or energy reductions (representing 59% of the FTSE 100). In June 2012, the UK coalition government announced the introduction of mandatory carbon reporting, requiring around 1,100 of the UK's largest listed companies to report their greenhouse gas emissions every year. Deputy Prime Minister Nick Clegg confirmed that emission reporting rules would come into effect from April 2013 in his piece for The Guardian.

Enterprise Carbon Accounting

Enterprise Carbon Accounting (ECA) or Corporate Carbon Footprint aims to be a rapid and cost effective process for businesses to collect, summarise, and report enterprise and supply chain GHG inventories. ECA leverages financial accounting principles, whilst utilising a hybrid of input-output LCA (Life Cycle Analysis)and process methodologies as appropriate. The evolution to ECA is necessary to address the urgent need for a more comprehensive and scalable approach to carbon accounting. While an emerging area, a number of new companies offer ECA solutions. ECA is a critical part of broader Enterprise Sustainability Accounting.

To be successful, an Enterprise Carbon Accounting system should have the following characteristics:

- Comprehensive: Incorporates Scope 1,2 and 3 emissions.

- Periodic: Enables updates at regular intervals and comparisons across reporting periods.

- Auditable: Traces transactions and enables independent reviews for compliance.

- Flexible: Incorporates data from multiple approaches to life cycle analysis.

- Standards-Based: Accommodates existing generally accepted standards and emerging standards.

- Scalable: Accommodates growing volume and complexity of business operations.

- Efficient: Delivers data in the timeframe required for decision making.

Enterprises that realize reduced emissions and energy consumption utilize systems with the following capabilities:

- Real-time historical energy data that is easily accessible.

- Role-based visibility into plant emissions data.

- Provides executives with real-time visibility into emissions data.

- Ability to benchmark emissions levels with goals and industry standards.

Life Cycle Analysis of ECA

Process LCA

Process LCA is the most popular method, currently, for conducting life-cycle assessment, and is often referred to as the SETAC-EPA method because of the role played by SETAC and EPA in this method's development. The inputs and outputs of multiple stages of a product's life are investigated in turn, and the results are aggregated into single metrics of impact such as eutrophication, toxicity, and greenhouse gas emissions. Three tools exist on the market to assist researchers in conducting process LCA (such as GaBi, Ecoinvent, and Umberto). These tools contain data from previous researchers on the environmental impact of materials and processes that are then strung together by the user to form a system.

Economic Input-output LCA

Input-Output LCA utilizes economic input-output tables and industry-level environmental data to construct a database of environmental impacts per dollar sold by an industry. The boundary problem of process LCA is solved in this method because the economic input-output table captures the interrelations of all economic sectors; however, aggregated industrial categories limit the specificity of the results. Input–output analysis is a very powerful tool for the upfront screening of corporate carbon footprints, for informing streamlined supply-chain GHG accounting and for setting priorities for more detailed analyses.

Hybrid LCA

Many methods for hybrid life-cycle assessments have been discussed, which aim to combine the infinite boundary of EIO-LCA with the specificity of Process LCA.

Enterprise Carbon Accounting (ECA)

At its core, ECA is essentially a hybrid life-cycle assessment; however, rather than the traditional bottom-up approach of life-cycle assessment, ECA links financial data directly to LCA data to produce a snapshot of the companies' operations. Rather than probing at areas thought to be problematic, ECA quickly identifies problem areas in the supply chain so that rapid action can be taken. This fundamental shift in thinking enables decision makers to rapidly address critical areas within the enterprise and supply chain.

Socialised Supply Chain

Socialised supply chain accounting is the term generally applied to Enterprise Carbon Accounting Solutions that provide a collaborative mechanism for supply chain participants to engage, expose and determine supply chain emissions through the process of shared knowledge. The term "Socialised Supply Chain" was coined by the CEO of Nootrol, Mark Kearns to describe a platform where supply chain participants exposed Process LCA and embedded emissions.

CARBON OFFSET

A carbon offset is a reduction in emissions of carbon dioxide or other greenhouse gases made in order to compensate for emissions made elsewhere. Offsets are measured in tonnes of carbon dioxide-equivalent (CO_2e). One tonne of carbon offset represents the reduction of one tonne of carbon dioxide or its equivalent in other greenhouse gases.

There are two markets for carbon offsets. In the larger, compliance market, companies, governments, or other entities buy carbon offsets in order to comply with caps on the total amount of carbon dioxide they are allowed to emit. For instance, an entity could be complying with obligations of Annex 1 Parties under the Kyoto Protocol or of liable entities under the EU Emission Trading Scheme, among others. In 2006, about $5.5 billion of carbon offsets were purchased in the compliance market, representing about 1.6 billion metric tons of CO_2e reductions.

In the much smaller, voluntary market, individuals, companies, or governments purchase carbon offsets to mitigate their own greenhouse gas emissions from transportation, electricity use, and other sources. For example, an individual might purchase carbon offsets to compensate for the greenhouse gas emissions caused by personal air travel. Carbon offset vendors offer direct purchase of carbon offsets, often also offering other services such as designating a carbon offset project to support or measuring a purchaser's carbon footprint. In 2016, about $191.3 million of carbon offsets were purchased in the voluntary market, representing about 63.4 million metric tons of CO_2e reductions.

Offsets typically support projects that reduce the emission of greenhouse gases in the short- or long-term. A common project type is renewable energy, such as wind farms, biomass energy, or hydroelectric dams. Others include energy efficiency projects, the destruction of industrial pollutants or agricultural byproducts, destruction of landfill methane, and forestry projects. Some of the most popular carbon offset projects from a corporate perspective are energy efficiency and wind turbine projects.

The Kyoto Protocol has sanctioned offsets as a way for governments and private companies to earn carbon credits that can be traded on a marketplace. The protocol established the Clean Development Mechanism (CDM), which validates and measures projects to ensure they produce authentic benefits and are genuinely "additional" activities that would not otherwise have been undertaken. Organizations that are unable to meet their emissions quota can offset their emissions by buying CDM-approved Certified Emissions Reductions.

Offsets may be cheaper or more convenient alternatives to reducing one's own fossil-fuel consumption. However, some critics object to carbon offsets, and question the benefits of certain types of offsets. Due diligence is recommended to help businesses in the assessment and identification of "good quality" offsets to ensure offsetting provides the desired additional environmental benefits, and to avoid reputational risk associated with poor quality offsets.

Offsets are viewed as an important policy tool to maintain stable economies and to improve sustainability. One of the hidden dangers of climate change policy is unequal prices of carbon in the economy, which can cause economic collateral damage if production flows to regions or industries that have a lower price of carbon—unless carbon can be purchased from that area, which offsets effectively permit, equalizing the price.

Wind turbines near Aalborg, Denmark. Renewable energy projects are the most common source of carbon offsets.

Features

Carbon offsets represent multiple categories of greenhouse gases, including carbon dioxide (CO_2), methane (CH_4), nitrous oxide (N_2O), perfluorocarbons (PFCs), hydrofluorocarbons (HFCs), and sulfur hexafluoride (SF_6).

Carbon offsets have several common features:

- Vintage: The vintage is the year in which the carbon emissions reduction takes place. Emissions reductions could be occurring in the future, meaning that the project developer anticipates future emissions, or could have already occurred, meaning that the purchaser is compensating the project developer for already-reduced emissions.

- Source: The source refers to the project or technology used in offsetting the carbon emissions. Projects can include land-use, methane, biomass, renewable energy and industrial energy efficiency. Projects may also have secondary benefits (co-benefits). For example, projects that reduce agricultural greenhouse gas emissions may improve water quality by reducing fertilizer usage.

- Certification regime: The certification regime describes the systems and procedures that are used to certify and register carbon offsets. Different methodologies are used for measuring and verifying emissions reductions, depending on project type, size and location. For example, the Clean Development Mechanism (CDM) differentiates between large and small scale projects. In the voluntary market, a variety of industry standards exist. These include the Verified Carbon Standard, Plan Vivo Foundation, and the Gold Standard, which are implemented to provide third-party verification of carbon offset projects. Gold Standard requires delivery and verification of sustainable development benefits alongside emission reductions. There are also some additional standards for the validation of co-benefits, including the CCBS, issued by Verra and the Social Carbon Standard, issued by the Ecologica Institute.

Types of Offset Projects

The CDM identifies over 200 types of projects suitable for generating carbon offsets, which are grouped into broad categories. These project types include renewable energy, methane abatement, energy efficiency, reforestation and fuel switching (i.e. to carbon-neutral fuels and carbon-negative fuels).

Renewable Energy

Renewable energy offsets commonly include wind power, solar power, hydroelectric power and biofuel. Some of these offsets are used to reduce the cost differential between renewable and conventional energy production, increasing the commercial viability of a choice to use renewable energy sources. Emissions from burning fuel, such as red diesel, has pushed one UK fuel supplier to create a carbon offset fuel named Carbon Offset Red Diesel.

Renewable Energy Credits (RECs) are also sometimes treated as carbon offsets, although the concepts are distinct. Whereas a carbon offset represents a reduction in greenhouse gas emissions, a REC represents a quantity of energy produced from renewable sources. To convert RECs into offsets, the clean energy must be translated into carbon reductions, typically by assuming that the clean energy is displacing an equivalent amount of conventionally produced electricity from the local grid. This is known as an indirect offset (because the reduction doesn't take place at the project site itself, but rather at an external site), and some controversy surrounds the question of whether they truly lead to "additional" emission reductions and who should get credit for any reductions that may occur. Intel corporation is the largest purchaser of renewable power in the US.

Methane Collection and Combustion

Some offset projects consist of the combustion or containment of methane generated by farm animals (by use of an anaerobic digester), landfills or other industrial waste. Methane has a global warming potential (GWP) 23 times that of CO_2; when combusted, each molecule of methane is converted to one molecule of CO_2, thus reducing the global warming effect by 96%.

An example of a project using an anaerobic digester can be found in Chile where in December 2000, the largest pork production company in Chile, initiated a voluntary process to implement advanced waste management systems (anaerobic and aerobic digestion of hog manure), in order to reduce greenhouse gas (GHG) emissions.

Energy Efficiency

While carbon offsets that fund renewable energy projects help lower the carbon intensity of energy supply, energy conservation projects seek to reduce the overall demand for energy. Carbon offsets in this category fund projects of several types:

Chicago Climate Justice activists protesting cap and trade
legislation in front of Chicago Climate Exchange building in Chicago Loop.

- Cogeneration plants generate both electricity and heat from the same power source, thus improving upon the energy efficiency of most power plants, which waste the energy generated as heat.

- Fuel efficiency projects replace a combustion device with one using less fuel per unit of energy provided. Assuming energy demand does not change, this reduces the carbon dioxide emitted.

- Energy-efficient buildings reduce the amount of energy wasted in buildings through efficient heating, cooling or lighting systems. In particular, the replacement of incandescent light bulbs with LED lamps can have a drastic effect on energy consumption. New buildings can also be constructed using less carbon-intensive input materials.

Destruction of Industrial Pollutants

Industrial pollutants such as hydrofluorocarbons (HFCs) and perfluorocarbons (PFCs) have a GWP many thousands of times greater than carbon dioxide by volume. Because these pollutants are easily captured and destroyed at their source, they present a large and low-cost source of carbon offsets. As a category, HFCs, PFCs, and N_2O reductions represent 71 per cent of offsets issued under the CDM.

Land use, Land-use Change and Forestry

Land use, land-use change and forestry (LULUCF) projects focus on natural carbon sinks such as forests and soil. Deforestation, particularly in Brazil, Indonesia and parts of Africa, account for about 20 per cent of greenhouse gas emissions. Deforestation can be avoided either by paying directly for forest preservation, or by using offset funds to provide substitutes for forest-based products. There is a class of mechanisms referred to as REDD schemes (Reducing emissions from deforestation and forest degradation), which may be included in a post-Kyoto agreement. REDD credits provide carbon offsets for the protection of forests, and provide a possible mechanism to allow funding from developed nations to assist in the protection of native forests in developing nations.

Almost half of the world's people burn wood (or fiber or dung) for their cooking and heating needs. Fuel-efficient cook stoves can reduce fuel wood consumption by 30 to 50%, though the warming of the earth due to decreases in particulate matter (i.e. smoke) from such fuel-efficient stoves has not been addressed. There are a number of different types of LULUCF projects:

- Avoided deforestation is the protection of existing forests.

- Reforestation is the process of restoring forests on land that was once forested.

- Afforestation is the process of creating forests on land that was previously un-forested, typically for longer than a generation.

- Soil management projects attempt to preserve or increase the amount of carbon sequestered in soil.

Links with Emission Trading Schemes

Once it has been accredited by the UNFCCC a carbon offset project can be used as carbon credit and linked with official emission trading schemes, such as the European Union Emission Trading Scheme or Kyoto Protocol, as Certified Emission Reductions. European emission allowances for the 2008–2012 second phase were selling for between 21 and 24 Euros per metric ton of CO_2 as of July 2007.

The voluntary Chicago Climate Exchange also includes a carbon offset scheme that allows offset project developers to sell emissions reductions to CCX members who have voluntarily agreed to meet emissions reduction targets.

The Western Climate Initiative, a regional greenhouse gas reduction initiative by states and provinces along the western rim of North America, includes an offset scheme. Likewise, the Regional Greenhouse Gas Initiative, a similar program in the northeastern U.S., includes an offset program. A credit mechanism that uses offsets may be incorporated in proposed schemes such as the Australian Carbon Exchange.

Carbon Retirement

Carbon retirement involves retiring allowances from emission trading schemes as a method for offsetting carbon emissions. Voluntary purchasers can offset their carbon emissions by purchasing carbon allowances from legally mandated cap-and-trade programs such as the Regional Greenhouse Gas Initiative or the European Emissions Trading Scheme. By purchasing the allowances that power plants, oil refineries, and industrial facilities need to hold to comply with a cap, voluntary purchases tighten the cap and force additional emissions reductions.

Small-scale Schemes

Voluntary purchases can also be made through small-scale and sometimes uncertified schemes such as those offered at South African based Promoting Access to Carbon Equity Centre (PACE), which nevertheless offer clear services such as poverty alleviation in the form of renewable energy development. These projects have the potential to develop projects that are either too small or too complicated to benefit from legally mandated cap-and-trade programs.

Other

A UK offset provider set up a carbon offsetting scheme that set up a secondary market for treadle pumps in developing countries. These pumps are used by farmers, using human power, in place of diesel pumps. However, given that treadle pumps are best suited to pumping shallow water, while diesel pumps are usually used to pump water from deep boreholes, it is not clear that the treadle pumps are actually achieving real emissions reductions. Other companies have explored and rejected treadle pumps as a viable carbon offsetting approach due to these concerns.

Accounting for and Verifying Reductions

Due to their indirect nature, many types of offset are difficult to verify. Some providers obtain independent certification that their offsets are accurately measured, to distance themselves from potentially fraudulent competitors. The credibility of the various certification providers is often questioned. Certified offsets may be purchased from commercial or non-profit organizations for US\$2.75–99.00 per tonne of CO_2, due to fluctuations of market price. Annual carbon dioxide emissions in developed countries range from 6 to 23 tons per capita.

Accounting systems differ on precisely what constitutes a valid offset for voluntary reduction systems and for mandatory reduction systems. However formal standards for quantification exist based on collaboration between emitters, regulators, environmentalists and project developers. These standards include the Voluntary Carbon Standard, Plan Vivo Foundation, Green-e Climate, Chicago Climate Exchange and the Gold Standard, the latter of which expands upon the requirements for the Clean Development Mechanism of the Kyoto Protocol.

Criteria for Quality Offsets

Accounting of offsets may address the following basic areas:

- Baseline and Measurement: What emissions would occur in the absence of a proposed project? And how are the emissions that occur after the project is performed going to be measured?

- Additionality: Would the project occur anyway without the investment raised by selling carbon offset credits? There are two common reasons why a project may lack additionality: (a) if it is intrinsically financially worthwhile due to energy cost savings, and (b) if it had to be performed due to environmental laws or regulations.

- Permanence: Are some benefits of the reductions reversible? (for example, trees may be harvested to burn the wood, and does growing trees for fuel wood decrease the need for fossil fuel?) If woodlands are increasing in area or density, then carbon is being sequestered. After roughly 50 years, newly planted forests will reach maturity and remove carbon dioxide more slowly.

- Leakage: Does implementing the project cause higher emissions outside the project boundary?

- Co-benefits: Are there other benefits in addition to the carbon emissions reduction, and to what degree?

Co-benefits

While the primary goal of carbon offsets is to reduce global carbon emissions, many offset projects also claim to lead to improvements in the quality of life for a local population. These additional improvements are termed co-benefits, and may be considered when evaluating and comparing carbon offset projects. For example, possible co-benefits from a project that replaces wood-burning stoves with ovens using a less carbon-intensive fuel could include:

- Lower non–greenhouse gas pollution (smoke, ash, and chemicals), which improves health in the home.

- Better preservation of forests, which are an important habitat for wildlife.

Offset projects can also lead to co-benefits such as better air and water quality, and healthier communities.

In a recent survey conducted by EcoSecurities, Conservation International, CCBA and ClimateBiz, of the 120 corporates surveyed more than 77 per cent rated community and environmental benefits as the prime motivator for purchasing carbon offsets.

Carbon offset projects can also negatively affect quality of life. For example, people who earn their livelihoods from collecting firewood and selling it to households could become unemployed if firewood is no longer used. A July 2007 paper from the Overseas Development Institute offers some indicators to be used in assessing the potential developmental impacts of voluntary carbon offset schemes:

- What potential does the project have for income generation?

- What effects might a project have on future changes in land use and could conflicts arise from this?

- Can small-scale producers engage in the scheme?

- What are the 'add on' benefits to the country—for example, will it assist capacity-building in local institutions?

Putting a price on carbon encourages innovation by providing funding for new ways to reduce greenhouse gases in many sectors. Carbon reduction goals drive the demand for offsets and carbon trading, encouraging the development of this new industry and offering opportunities for different sectors to develop and use innovative new technologies.

Carbon offset projects also provide savings: Energy efficiency measures may reduce fuel or electricity consumption, leading to a potential reduction in maintenance and operating costs.

The UNFCCC has created a dedicated website where CDM activities and prior consideration projects are able to report their co-benefits on a voluntary basis.

CARBON SEQUESTRATION

Carbon sequestration is the process involved in carbon capture and the long-term storage of atmospheric carbon dioxide or other forms of carbon to mitigate or defer global warming. It has been proposed as a way to slow the atmospheric and marine accumulation of greenhouse gases, which are released by burning fossil fuels.

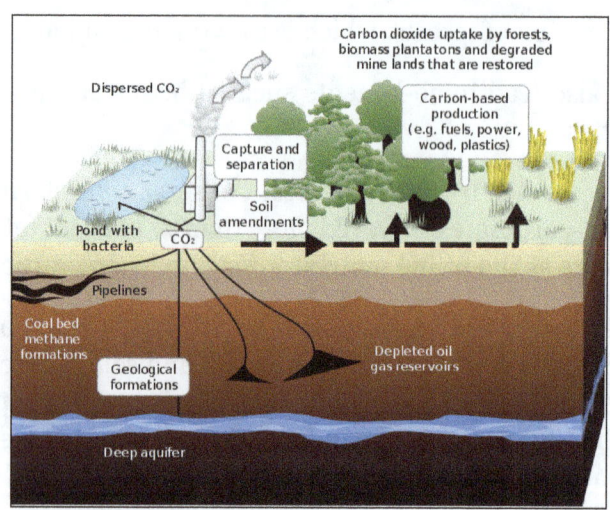

Schematic showing both terrestrial and geological
sequestration of carbon dioxide emissions from a coal-fired plant.

Carbon dioxide (CO_2) is naturally captured from the atmosphere through biological, chemical, and physical processes. Artificial processes have been devised to produce similar effects, including large-scale, artificial capture and sequestration of industrially produced CO_2 using subsurface saline aquifers, reservoirs, ocean water, aging oil fields, or other carbon sinks.

Biological Processes

An oceanic phytoplankton bloom in the South Atlantic Ocean,
off the coast of Argentina. Encouraging such blooms with
iron fertilization could lock up carbon on the seabed.

Biosequestration or carbon sequestration through biological processes affects the global carbon cycle. Examples include major climatic fluctuations, such as the Azolla event, which created the current Arctic climate. Such processes created fossil fuels, as well as clathrate and limestone. By manipulating such processes, geoengineers seek to enhance sequestration.

Peat Production

Peat bogs act as a sink for carbon due to the accumulation of partially decayed biomass that would otherwise continue to decay completely. There is a variance on how much the peatlands act as a carbon sink or carbon source that can be linked to varying climates in different areas of the world and different times of the year. By creating new bogs, or enhancing existing ones, the amount of carbon that is sequestered by bogs would increase.

Forestry

Afforestation is the establishment of a forest in an area where there was no previous tree cover. Reforestation is the replanting of trees on marginal crop and pasture lands to incorporate carbon from atmospheric CO_2 into biomass. For this carbon sequestration process to succeed the carbon must not return to the atmosphere from mass burning or rotting when the trees die. To this end, land allotted to the trees must not be converted to other uses and management of the frequency of disturbances might be necessary in order to avoid extreme events. Alternatively, the wood from them must itself be sequestered, e.g., via biochar, bio-energy with carbon storage (BECS), landfill or 'stored' by use in e.g. construction. Short of growth in perpetuity, however, reforestation with long-lived trees (>100 years) will sequester carbon for a more graduated release, minimizing impact during the expected carbon crisis of the 21st century. According to a research by Tom Crowther et al, there is still enough room to plant an additional 1.2 trillion trees. This amount of trees would cancel out the last 10 years of CO_2 emissions and sequester 160 billion tons of carbon. This vision is being executed by the Trillion Tree Campaign. According to research, restoring all degraded forests all over the world could capture about 205 billion tons of carbon in total (which is about 2/3rd of all carbon emissions, bringing global warming down to below 2°C).

Urban Forestry

Urban forestry increases the amount of carbon taken up in cities by adding new tree sites and the sequestration of carbon occurs over the lifetime of the tree. It is generally practiced and maintained on smaller scales, like in cities. The results of urban forestry can have different results depending on the type of vegetation that is being used, so it can function as a sink but can also function as a source of emissions. Along with sequestration by the plants which is difficult to measure but seems to have little effect on the overall amount of carbon dioxide that is uptaken, the vegetation can have indirect effects on carbon by reducing need for energy consumption.

Seaweed Farming

Large-scale seaweed farming (called "ocean afforestation") could sequester huge amounts of carbon. Afforesting just 9% of the ocean could sequester 53 billion tons of carbon dioxide annually.

Wetland Restoration

Wetland soil is an important carbon sink; 14.5% of the world's soil carbon is found in wetlands, while only 6% of the world's land is composed of wetlands.

Agriculture

Compared to natural vegetation, cropland soils are depleted in soil organic carbon (SOC). When a soil is converted from natural land or semi natural land, such as forests, woodlands, grasslands, steppes and savannas, the SOC content in the soil reduces with about 30–40%. This loss is due to the removal of plant material containing carbon, in terms of harvests. When the land use changes, the carbon in the soil will either increase or decrease, this change will continue until the soil reaches a new equilibrium. Deviations from this equilibrium can also be affected by variated climate. The decreasing of SOC content can be counteracted by increasing the carbon input, this can be done with several strategies, e.g. leave harvest residues on the field, use manure as fertiliser or include perennial crops in the rotation. Perennial crops have larger below ground biomass fraction, which increases the SOC content. Globally, soils are estimated to contain approximately 1,500 gigatons of organic carbon to 1 m depth, more than the amount in vegetation and the atmosphere.

Modification of agricultural practices is a recognized method of carbon sequestration as soil can act as an effective carbon sink offsetting as much as 20% of 2010 carbon dioxide emissions annually.

Carbon emission reduction methods in agriculture can be grouped into two categories: reducing and displacing emissions and enhancing carbon removal. Some of these reductions involve increasing the efficiency of farm operations (e.g. more fuel-efficient equipment) while some involve interruptions in the natural carbon cycle. Also, some effective techniques (such as the elimination of stubble burning) can negatively impact other environmental concerns (increased herbicide use to control weeds not destroyed by burning).

Deep Soil

Soils hold four times the amount of carbon stored in the atmosphere. About half of this is found deep within soils. About 90% of this deep soil C is stabilized by mineral-organic associations.

Reducing Emissions

Increasing yields and efficiency generally reduces emissions as well, since more food results from the same or less effort. Techniques include more accurate use of fertilizers, less soil disturbance, better irrigation, and crop strains bred for locally beneficial traits and increased yields.

Replacing more energy intensive farming operations can also reduce emissions. Reduced or no-till farming requires less machine use and burns correspondingly less fuel per acre. However, no-till usually increases use of weed-control chemicals and the residue now left on the soil surface is more likely to release its CO_2 to the atmosphere as it decays, reducing the net carbon reduction.

In practice, most farming operations that incorporate post-harvest crop residues, wastes and byproducts back into the soil provide a carbon storage benefit. This is particularly the case for practices such as field burning of stubble – rather than releasing almost all of the stored CO_2 to the atmosphere, tillage incorporates the biomass back into the soil.

Enhancing Carbon Removal

All crops absorb CO_2 during growth and release it after harvest. The goal of agricultural carbon removal is to use the crop and its relation to the carbon cycle to permanently sequester carbon within the soil. This is done by selecting farming methods that return biomass to the soil and enhance the conditions in which the carbon within the plants will be reduced to its elemental nature and stored in a stable state. Methods for accomplishing this include:

- Use cover crops such as grasses and weeds as temporary cover between planting seasons.

- Concentrate livestock in small paddocks for days at a time so they graze lightly but evenly. This encourages roots to grow deeper into the soil. Stock also till the soil with their hooves, grinding old grass and manures into the soil.

- Cover bare paddocks with hay or dead vegetation. This protects soil from the sun and allows the soil to hold more water and be more attractive to carbon-capturing microbes.

- Restore degraded land, which slows carbon release while returning the land to agriculture or other use.

Agricultural sequestration practices may have positive effects on soil, air, and water quality, be beneficial to wildlife, and expand food production. On degraded croplands, an increase of 1 ton of soil carbon pool may increase crop yield by 20 to 40 kilograms per hectare of wheat, 10 to 20 kg/ ha for maize, and 0.5 to 1 kg/ha for cowpeas.

The effects of soil sequestration can be reversed. If the soil is disrupted or tillage practices are abandoned, the soil becomes a net source of greenhouse gases. Typically after 15 to 30 years of sequestration, soil becomes saturated and ceases to absorb carbon. This implies that there is a global limit to the amount of carbon that soil can hold.

Many factors affect the costs of carbon sequestration including soil quality, transaction costs and various externalities such as leakage and unforeseen environmental damage. Because reduction of atmospheric CO_2 is a long-term concern, farmers can be reluctant to adopt more expensive agricultural techniques when there is not a clear crop, soil, or economic benefit. Governments such as Australia and New Zealand are considering allowing farmers to sell carbon credits once they document that they have sufficiently increased soil carbon content.

Ocean-related

Iron Fertilization

Ocean iron fertilization is an example of such a geoengineering technique. Iron fertilization attempts to encourage phytoplankton growth, which removes carbon from the atmosphere for at least a period of time. This technique is controversial due to limited understanding of its complete effects on the marine ecosystem, including side effects and possibly large deviations from expected behavior. Such effects potentially include release of nitrogen oxides, and disruption of the ocean's nutrient balance.

Natural iron fertilisation events (e.g., deposition of iron-rich dust into ocean waters) can enhance carbon sequestration. Sperm whales act as agents of iron fertilisation when they transport iron from the deep ocean to the surface during prey consumption and defecation. Sperm whales have been shown to increase the levels of primary production and carbon export to the deep ocean by depositing iron rich feces into surface waters of the Southern Ocean. The iron rich feces causes phytoplankton to grow and take up more carbon from the atmosphere. When the phytoplankton dies, some of it sinks to the deep ocean and takes the atmospheric carbon with it. By reducing the abundance of sperm whales in the Southern Ocean, whaling has resulted in an extra 200,000 tonnes of carbon remaining in the atmosphere each year.

Urea Fertilization

Ian Jones proposes fertilizing the ocean with urea, a nitrogen rich substance, to encourage phytoplankton growth.

Australian company Ocean Nourishment Corporation (ONC) plans to sink hundreds of tonnes of urea into the ocean to boost CO_2-absorbing phytoplankton growth as a way to combat climate change. In 2007, Sydney-based ONC completed an experiment involving 1 tonne of nitrogen in the Sulu Sea off the Philippines.

Mixing Layers

Encouraging various ocean layers to mix can move nutrients and dissolved gases around, offering avenues for geoengineering. Mixing may be achieved by placing large vertical pipes in the oceans to pump nutrient rich water to the surface, triggering blooms of algae, which store carbon when they grow and export carbon when they die. This produces results somewhat similar to iron fertilization. One side-effect is a short-term rise in CO_2, which limits its attractiveness.

Seaweed

Seaweed grows very fast and can theoretically be harvested and processed to generate biomethane, via Anaerobic Digestion to generate electricity, via Cogeneration/CHP or as a replacement for natural gas. One study suggested that if seaweed farms covered 9% of the ocean they could produce enough biomethane to supply Earth's equivalent demand for fossil fuel energy, remove 53 gigatonnes of CO_2 per year from the atmosphere and sustainably produce 200 kg per year of fish, per person, for 10 billion people. Ideal species for such farming and conversion include Laminaria digitata, Fucus serratus and Saccharina latissima.

Physical Processes

Biochar can be landfilled, used as a soil improver
or burned using carbon capture and storage.

Bio-energy with Carbon Capture and Storage

Bio-energy with carbon capture and storage (BECCS) refers to biomass in power stations and boilers that use carbon capture and storage. The carbon sequestered by the biomass would be captured and stored, thus removing carbon dioxide from the atmosphere.

This technology is sometimes referred to as bio-energy with carbon storage, BECS, though this term can also refer to the carbon sequestration potential in other technologies, such as biochar.

Burial

Burying biomass (such as trees) directly, mimics the natural processes that created fossil fuels. Landfills also represent a physical method of sequestration.

Biochar Burial

Biochar is charcoal created by pyrolysis of biomass waste. The resulting material is added to a landfill or used as a soil improver to create terra preta. Addition of pyrogenic organic carbon (biochar) is a novel strategy to increase the soil-C stock for the long-term and to mitigate global-warming by offsetting the atmospheric C (up to 9.5 Pg C annually).

In the soil, the carbon is unavailable for oxidation to CO_2 and consequential atmospheric release. This is one technique advocated by scientist James Lovelock, creator of the Gaia hypothesis. According to Simon Shackley, "people are talking more about something in the range of one to two billion tonnes a year."

The mechanisms related to biochar are referred to as bio-energy with carbon storage, BECS.

Ocean Storage

If CO_2 were to be injected to the ocean bottom, the pressures would be great enough for CO_2 to be in its liquid phase. The idea behind ocean injection would be to have stable, stationary pools of CO_2 at the ocean floor. The ocean could potentially hold over a thousand billion tons of CO_2. However, this avenue of sequestration isn't being as actively pursued because of concerns about the impact on ocean life, and concerns about its stability.

River mouths bring large quantities of nutrients and dead material from upriver into the ocean as part of the process that eventually produces fossil fuels. Transporting material such as crop waste out to sea and allowing it to sink exploits this idea to increase carbon storage. International regulations on marine dumping may restrict or prevent use of this technique.

Geological Sequestration

Geological sequestration refers to the storage of CO_2 underground in depleted oil and gas reservoirs, saline formations, or deep, un-minable coal beds.

Once CO_2 is captured from a gas or coal-fired power plant, it would be compressed to ≈100 bar so that it would be a supercritical fluid. In this fluid form, the CO_2 would be easy to transport via pipeline to the place of storage. The CO_2 would then be injected deep underground, typically around 1 km, where it would be stable for hundreds to millions of years. At these storage conditions, the density of supercritical CO_2 is 600 to 800 kg/m³. For consumers, the cost of electricity from a coal-fired power plant with carbon capture and storage (CCS) is estimated to be 0.01–0.05 \$/kWh higher than without CCS. For reference, the average cost of electricity in the US in 2004 was 0.0762 \$/kWh. In other terms, the cost of CCS would be 20–70 \$/ton of CO_2 captured. The transportation and injection of CO_2 is relatively cheap, with the capture costs accounting for 70–80% of CCS costs.

The important parameters in determining a good site for carbon storage are: rock porosity, rock permeability, absence of faults, and geometry of rock layers. The medium in which the CO_2 is to be stored ideally has a high porosity and permeability, such as sandstone or limestone. Sandstone can have a permeability ranging from 1 to 10^{-5} Darcy, and can have a porosity as high as $\approx 30\%$. The porous rock must be capped by a layer of low permeability which acts as a seal, or caprock, for the CO_2. Shale is an example of a very good caprock, with a permeability of 10^{-5} to 10^{-9} Darcy. Once injected, the CO_2 plume will rise via buoyant forces, since it is less dense than its surroundings. Once it encounters a caprock, it will spread laterally until it encounters a gap. If there are fault planes near the injection zone, there is a possibility the CO_2 could migrate along the fault to the surface, leaking into the atmosphere, which would be potentially dangerous to life in the surrounding area. Another danger related to carbon sequestration is induced seismicity. If the injection of CO_2 creates pressures that are too high underground, the formation will fracture, causing an earthquake.

While trapped in a rock formation, CO_2 can be in the supercritical fluid phase or dissolve in groundwater/brine. It can also react with minerals in the geologic formation to precipitate carbonates.

Worldwide storage capacity in oil and gas reservoirs is estimated to be 675–900 Gt CO_2, and in un-minable coal seams is estimated to be 15–200 Gt CO_2. Deep saline formations have the largest capacity, which is estimated to be 1,000–10,000 Gt CO_2. In the US, there is an estimated 160 Gt CO_2 storage capacity.

There are a number of large-scale carbon capture and sequestration projects that have demonstrated the viability and safety of this method of carbon storage, which are summarized here by the Global CCS Institute. The dominant monitoring technique is seismic imaging, where vibrations are generated that propagate through the subsurface. The geologic structure can be imaged from the refracted/reflected waves.

The first large-scale CO_2 sequestration project which began in 1996 is called Sleipner, and is located in the North Sea where Norway's StatoilHydro strips carbon dioxide from natural gas with amine solvents and disposed of this carbon dioxide in a deep saline aquifer. In 2000, a coal-fueled synthetic natural gas plant in Beulah, North Dakota, became the world's first coal-using plant to capture and store carbon dioxide, at the Weyburn-Midale Carbon Dioxide Project.

CO_2 has been used extensively in enhanced crude oil recovery operations in the United States beginning in 1972. There are in excess of 10,000 wells that inject CO_2 in the state of Texas alone. The gas comes in part from anthropogenic sources, but is principally from large naturally occurring geologic formations of CO_2. It is transported to the oil-producing fields through a large network of over 5,000 kilometres (3,100 mi) of CO_2 pipelines. The use of CO_2 for enhanced oil recovery (EOR) methods in heavy oil reservoirs in the Western Canadian Sedimentary Basin (WCSB) has also been proposed. However, transport cost remains an important hurdle. An extensive CO_2 pipeline system does not yet exist in the WCSB. Athabasca oil sands mining that produces CO_2 is hundreds of kilometers north of the subsurface Heavy crude oil reservoirs that could most benefit from CO_2 injection.

Chemical Processes

Developed in the Netherlands, an electrocatalysis by a copper complex helps reduce carbon dioxide to oxalic acid; this conversion uses carbon dioxide as a feedstock to generate oxalic acid.

Mineral Carbonation

Carbon, in the form of CO_2 can be removed from the atmosphere by chemical processes, and stored in stable carbonate mineral forms. This process is known as 'carbon sequestration by mineral carbonation' or mineral sequestration. The process involves reacting carbon dioxide with abundantly available metal oxides–either magnesium oxide (MgO) or calcium oxide (CaO)–to form stable carbonates. These reactions are exothermic and occur naturally (e.g., the weathering of rock over geologic time periods).

$$CaO + CO_2 \rightarrow CaCO_3$$

$$MgO + CO_2 \rightarrow MgCO_3$$

Calcium and magnesium are found in nature typically as calcium and magnesium silicates (such as forsterite and serpentinite) and not as binary oxides. For forsterite and serpentine the reactions are:

$$Mg_2SiO_4 + 2\ CO_2 \rightarrow 2\ MgCO_3 + SiO_2$$

$$Mg_3Si_2O_5(OH)_4 + 3\ CO_2 \rightarrow 3\ MgCO_3 + 2\ SiO_2 + 2\ H_2O$$

Theoretically up to 22% of this mineral mass is able to form carbonates.

Earthen Oxide	Percent of Crust	Carbonate	Enthalpy change (kJ/mol)
SiO_2	59.71		
Al_2O_3	15.41		
CaO	4.90	$CaCO_3$	−179
MgO	4.36	$MgCO_3$	−117
Na_2O	3.55	Na_2CO_3	
FeO	3.52	$FeCO_3$	
K_2O	2.80	K_2CO_3	
Fe_2O_3	2.63	$FeCO_3$	
	21.76	All Carbonates	

These reactions are slightly more favorable at low temperatures. This process occurs naturally over geologic time frames and is responsible for much of the Earth's surface limestone. The reaction rate can be made faster however, by reacting at higher temperatures and pressures, although this method requires some additional energy. Alternatively, the mineral could be milled to increase its surface area, and exposed to water and constant abrasion to remove the inert Silica as could be achieved naturally by dumping Olivine in the high energy surf of beaches. Experiments suggest the weathering process is reasonably quick (one year) given porous basaltic rocks.

CO_2 naturally reacts with peridotite rock in surface exposures of ophiolites, notably in Oman. It has been suggested that this process can be enhanced to carry out natural mineralisation of CO_2.

When CO_2 is dissolved in water and injected into hot basaltic rocks underground, it has been shown that the CO_2 reacts with the basalt to form solid carbonate minerals. A test plant in Iceland started up in October 2017, extracting up to 50 tons of CO_2 a year from the atmosphere and storing it underground in basaltic rock.

Researchers from British Columbia, developed a low cost process for the production of magnesite, also known as magnesium carbonate, which can sequester CO_2 from the air, or at the point of air pollution, e.g. at a power plant. The crystals are naturally occurring, but accumulation is usually very slow.

Electrochemical Method

Another method uses a liquid metal catalyst and an electrolyte liquid into which CO_2 is dissolved. The CO_2 then converts into solid flakes of carbon. This method is done at room temperature.

Industrial Use

Traditional cement manufacture releases large amounts of carbon dioxide, but newly developed cement types from Novacem can absorb CO_2 from ambient air during hardening. A similar technique was pioneered by TecEco, which has been producing "Eco-Cement" since 2002. A Canadian startup CarbonCure takes captured CO_2 and injects it into concrete as it is being mixed. Carbon Upcycling UCLA is another company that uses CO_2 in concrete. Their concrete product is called CO_2 NCRETE, a concrete that hardens faster and is more eco-friendly than traditional concrete.

In Estonia, oil shale ash, generated by power stations could be used as sorbents for CO_2 mineral sequestration. The amount of CO_2 captured averaged 60 to 65% of the carbonaceous CO_2 and 10 to 11% of the total CO_2 emissions.

Chemical Scrubbers

Various carbon dioxide scrubbing processes have been proposed to remove CO_2 from the air, usually using a variant of the Kraft process. Carbon dioxide scrubbing variants exist based on potassium carbonate, which can be used to create liquid fuels, or on sodium hydroxide. These notably include artificial trees proposed by Klaus Lackner to remove carbon dioxide from the atmosphere using chemical scrubbers.

Ocean-related

Basalt Storage

Carbon dioxide sequestration in basalt involves the injecting of CO_2 into deep-sea formations. The CO_2 first mixes with seawater and then reacts with the basalt, both of which are alkaline-rich elements. This reaction results in the release of Ca^{2+} and Mg^{2+} ions forming stable carbonate minerals.

Underwater basalt offers a good alternative to other forms of oceanic carbon storage because it has a number of trapping measures to ensure added protection against leakage. These measures include "geochemical, sediment, gravitational and hydrate formation." Because CO_2 hydrate is denser than CO_2 in seawater, the risk of leakage is minimal. Injecting the CO_2 at depths greater than 2,700 meters (8,900 ft) ensures that the CO_2 has a greater density than seawater, causing it to sink.

One possible injection site is Juan de Fuca plate. Researchers found that this plate at the western coast of the United States has a possible storage capacity of 208 gigatons. This could cover the entire current U.S. carbon emissions for over 100 years.

This process is undergoing tests as part of the CarbFix project, resulting in 95% of the injected 250 tonnes of CO_2 to solidify into calcite in 2 years, using 25 tonnes of water per tonne of CO_2.

Acid Neutralisation

Carbon dioxide forms carbonic acid when dissolved in water, so ocean acidification is a significant consequence of elevated carbon dioxide levels, and limits the rate at which it can be absorbed into the ocean (the solubility pump). A variety of different bases have been suggested that could neutralize the acid and thus increase CO_2 absorption. For example, adding crushed limestone to oceans enhances the absorption of carbon dioxide. Another approach is to add sodium hydroxide to oceans which is produced by electrolysis of salt water or brine, while eliminating the waste hydrochloric acid by reaction with a volcanic silicate rock such as enstatite, effectively increasing the rate of natural weathering of these rocks to restore ocean pH.

Obstruction

Danger of Leaks

Carbon dioxide may be stored deep underground. At depth, hydrostatic pressure acts to keep it in a liquid state. Reservoir design faults, rock fissures and tectonic processes may act to release the gas stored into the ocean or atmosphere.

Financial Costs

The use of the technology would add an additional 1–5 cents of cost per kilowatt hour, according to estimate made by the Intergovernmental Panel on Climate Change. The financial costs of modern coal technology would nearly double if use of CCS technology were to be required by regulation. The cost of CCS technology differs with the different types of capture technologies being used and with the different sites that it is implemented in, but the costs tend to increase with CCS capture implementation. One study conducted predicted that with new technologies these costs could be lowered but would remain slightly higher than prices without CCS technologies.

Energy Requirements

The energy requirements of sequestration processes may be significant. In one paper, sequestration consumed 25 percent of the plant's rated 600 megawatt output capacity.

After adding CO_2 capture and compression, the capacity of the coal-fired power plant is reduced to 457 MW.

INDIVIDUAL ACTION ON CLIMATE CHANGE

Individual action on climate change can include personal choices in many areas, such as diet, means of long- and short-distance travel, household energy use, consumption of goods and services, and family size. Individuals can also engage in local and political advocacy around issues of climate change.

A demonstrator at the People's Climate March.

As of 2019, emissions budgets are uncertain and estimates of the annual average carbon footprint per person required to meet climate change targets vary between 1 and 3 tonnes CO_2-equivalent, down from a 2018 world average of about 5 tonnes.

The IPCC Fifth Assessment Report emphasises that behavior, lifestyle and cultural change have a high mitigation potential in some sectors, particularly when complementing technological and structural change. In general, higher consumption lifestyles have a greater environmental impact. Several scientific studies have shown that when people, especially those living in developed countries but more generally including all countries, wish to reduce their carbon footprint, there are four key "high-impact" actions they can take:

- Not having an additional child ("an average for developed countries of 58.6 tonnes CO_2-equivalent (tCO_2e) emission reductions per year.")

- Living car-free (2.4 tonnes).

- Avoiding one round-trip transatlantic flight (1.6 tonnes).

- Eating a plant-based diet (0.8 tonnes).

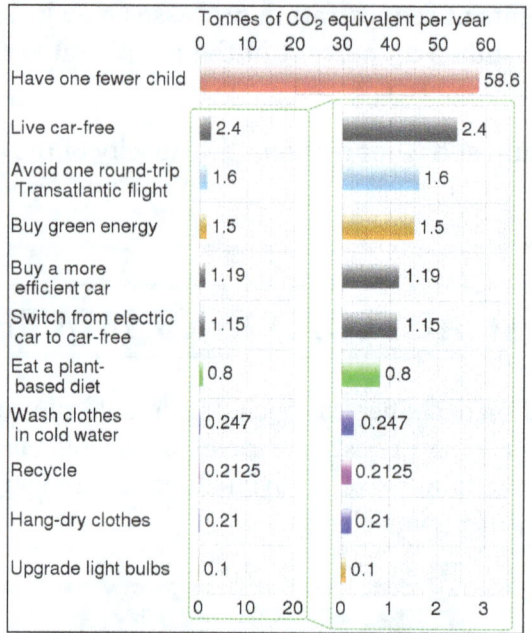

Reduction of one's carbon footprint for various actions.

These differ significantly from much popular advice for "greening" one's lifestyle, which seem to fall mostly into the "low-impact" category: Replacing a typical car with a hybrid (0.52 tonnes); washing clothes in cold water (0.25 tonnes); recycling (0.21 tonnes); upgrading light bulbs (0.10 tonnes); etc. The researchers found that public discourse on reducing one's carbon footprint overwhelmingly focuses on low-impact behaviors, and that mention of the high-impact behaviors is almost non-existent in the mainstream media, government publications, K-12 school textbooks, etc. The researchers added that "Our recommended high-impact actions are more effective than many more commonly discussed options (e.g. eating a plant-based diet saves eight times more emissions than upgrading light bulbs)."

Some commentators have argued that individual actions as consumers and "greening personal lives" are insignificant to collective action to hold fossil fuel corporations accountable, as they have produced 71% of carbon emissions since 1988. Others say that individual action leads to collective action.

Family Size

Although having fewer children is arguably the individual action that most effectively reduces a person's climate impact, the issue is rarely raised, and it is arguably controversial due to its private nature. Even so, ethicists, some politicians such as Alexandria Ocasio-Cortez, and others have started discussing the climate implications associated with reproduction.

It has been claimed that "a US family who chooses to have one fewer child would provide the same level of emissions reductions as 684 teenagers who choose to adopt comprehensive recycling for the rest of their lives." This has been criticised: both as a category mistake for assigning descendants emissions to their ancestors and for the very long timescale of reductions.

Two interrelated aspects of this action, family planning and women and girl's education, are modeled by Project Drawdown as the 6 and 7 top potential solutions for climate change, based on the ability of family planning and education to reduce the growth of the overall global population.

Travel and Commuting

In the early 21st century perception towards climate change influenced some people in rich countries to change their travel lifestyle.

Air Transport

Avoiding air travel and particularly frequent flyer programs has a high benefit because the convenience makes frequent, long distance travel easy, and high-altitude emissions are more potent for the climate than the same emissions made at ground level. Aviation is much more difficult to fix technically than surface transport, so will need more individual action in future if the Carbon Offsetting and Reduction Scheme for International Aviation cannot be made to work properly.

Surface Transport

- Walking and running are among the least environmentally harmful modes of transportation, followed by cycling.

- Public transport such as electric buses, metro and electric trains generally emit less greenhouse gases than cars.

- Electric kick scooters could also be a low-impact form of transportation, with emerging startups such as Bird and Lime providing shared scooters allowing for last-mile transportation. However, their short lifespan caused by rough usage and vandalism could mean additional resources spent on replacement units. Some models provide higher range (35+ miles, 56+ km) and speed (40+ mph, 64+ km/h), which can be used in areas with poor public transportation infrastructure where cars and motorcycles would have previously been the only option.

- Cars: Using an electric car instead of a gasoline or diesel car helps to reduce carbon dioxide emissions.

Diet and Food

Agriculture is very difficult to fix technically so will need more individual action or carbon offsetting than all other sectors except perhaps aviation.

Eating less meat, especially beef and lamb, reduces emissions. In 2019, the IPCC released a summary of the 2019 special report which asserted that a shift towards plant-based diets would help to mitigate and adapt to climate change. Ecologist Hans-Otto Pörtner, who contributed to the report, said "We don't want to tell people what to eat, but it would indeed be beneficial, for both climate and human health, if people in many rich countries consumed less meat, and if politics would create appropriate incentives to that effect."

Eating a plant-rich diet is listed as the 4 solution for climate change as modeled by Project Drawdown, based on avoided emissions from the production of animals and avoided emissions from additional deforestation for grazing land.

Home Energy, Landscaping and Consumption

Reducing home energy use through measures such as insulation, better energy efficiency of appliances, and improving heating and cooling efficiency can significantly reduce individual's carbon footprints.

In addition, the choice of fuel used to heat, cool, and power homes makes a difference in the carbon footprint of individual homes. Many energy suppliers in various countries worldwide have options to purchase part or pure "green energy" (usually electricity but occasionally also gas). These methods of energy production emit almost no greenhouse gases once they are up and running.

Installing rooftop solar, both on a household and community scale, also drastically reduces household emissions, and at scale could be a major contributor to greenhouse gas abatement.

Low Energy Products and Consumption

Labels, such as Energy Star in the US, can be seen on many household appliances, home electronics, office equipment, heating and cooling equipment, windows, residential light fixtures, and other products. These may help consumers choose lower energy products.

Landscape and Gardens

Protecting forests and planting new trees contributes to the absorption of carbon dioxide from the air. There are many opportunities to plant trees in the yard, along roads, in parks, and in public gardens. In addition, some charities plant fast-growing trees—for as little as $US0.10 per tree—to help people in tropical developing countries restore the productivity of their lands. Conversely, clearing old-growth forests adds to the carbon in the atmosphere, so buying non-old-growth paper is good for the climate as well as the forest.

Additionally, turfgrass lawns can contribute to climate change through the impact of using fertilizers, herbicides, irrigation, and gas-powered lawnmowers and other tools; depending on how lawns are managed, the impact of emissions from maintenance and chemicals may outweigh any carbon sequestration from the lawn. Reducing irrigation, chemical use, planting native plants or bushes, and using hand tools can all reduce the climate impact of lawns.

Individual Purchase of Carbon Offsets

The principle of carbon offset is thus: one decides that they don't want to be responsible for accelerating climate change, and they've already made efforts to reduce their carbon dioxide emissions, so they decide to pay someone else to further reduce their net emissions by planting trees or by taking up low-carbon technologies. Every unit of carbon that is absorbed by trees—or not emitted due to your funding of renewable energy deployment—offsets the emissions from their fossil fuel use. In many cases, funding of renewable energy, energy efficiency, or tree planting — particularly in developing nations—can be a relatively cheap way of making an individual "carbon neutral".

Citizen Participation in Climate Change Policy Advocacy

Some posit that citizen participation in groups advocating for collective action in the form of political solutions, such as carbon pricing, meat pricing, ending subsidies for fossil fuels and animal husbandry, and ending laws mandating car use, is the most impactful way for individuals to act to prevent climate change. It is argued that climate change is a collective action problem, specifically a tragedy of the commons, which is a political and not individual category of problem.

INDUSTRIAL ACTION ON CLIMATE CHANGE

The Industry sector produces goods and raw materials for everyday use, every single day. The greenhouse gases that industrial production emits are split into two categories:

- Direct Emissions: Greenhouse emissions that are produced at the facility itself.

- Indirect Emissions: Are associated with the facility's use of energy, but happens off-site.

Even relaxing comfort standards by turning down the heat while at night and keeping temperatures moderate at all times. Setting the thermostat just 2 degrees lower in winter and higher in summer could save about 2,000 pounds of carbon dioxide emissions each year.

Ways that the industry sector can reduce CO_2 emissions include:

- Measuring Carbon Footprint: By assessing how much pollution an organization's actions generate, you can begin to see how changing a few policies here, and there can significantly reduce the overall carbon footprint. A carbon footprint can be measured by undertaking a Greenhouse Gas (GHG) emissions assessment. Once the size of a carbon footprint is known, a strategy can be devised to reduce it, e.g., by technological developments, better process and product management, changed Green Public or Private Procurement (GPP), carbon capture, consumption strategies, and others.

- Carbon Capping: Emissions trading, sometimes known as cap-and-trade policies, puts a limit on carbon dioxide emissions. A government entity sets a "cap" on the emissions that can be produced in its jurisdiction, and companies are given carbon allowances. These allowances can either be used or traded to other companies

- Reducing Energy Use: The building industry now has more energy efficiency certifications than ever. The standards help set measurable and achievable goals for energy use reductions.

- The industry sector can ensure new buildings are made to be energy efficient by earning any of these ratings. Each of the rating systems assists building owners in reducing the amount of energy used from 12% all the way to 100% reduction in typical building energy use.

- Rewarding Green Commutes: Encouraging employees to switch to public transportation, carpooling, biking, telecommuting and other innovative ways to save energy and reduce greenhouse gas emissions on the way to and from work can add up and have tremendous effects. Employers can offer commuter benefits that address limited or expensive parking, reduce traffic congestion, improve employee recruiting and retention and minimize the environmental impacts associated with drive-alone commuting.

Learning to Adapt to Climate Change

Climate change is already being felt in towns and cities across the country. Hundreds of municipalities have centered their climate change efforts on mitigation work and have successfully reduced their greenhouse gas emissions and lessened their climate impacts. However, with the increasing effects of climate change becoming apparent, municipalities are beginning to assess their vulnerability to the changes that are already underway, and developing responses that protect their citizens and their economies.

ZERO CARBON BUILDINGS

ZCBs are buildings with a net zero amount of carbon emissions associated with their annual energy demand. ZCBs achieve this by:

- Implementing high levels of energy efficiency.

- Meeting energy needs with on- or off-site renewable energy sourcing.

In some cases, as a last resort, buildings can also partly achieve net zero emissions through carbon offsets, which often come in the form of renewable energy investments elsewhere. Offsets, however, are only recommended for cases in which a 100% renewable energy supply is not feasible.

While in the past ZCBs have been seen as a target only wealthy countries could reach, there are policy pathways today to reach zero carbon buildings regardless of location or development status.

We have the technology to achieve ZCBs in nearly every context; what national and local leaders need are policy pathways and financing solutions. There are multiple ways to achieve ZCBs through combinations of energy efficiency, renewable energy and carbon offsets, in the following order of priority:

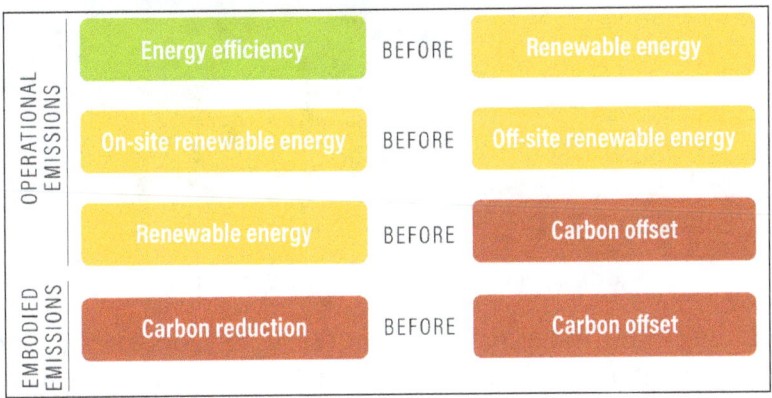

The more carbon avoided through efficiency, the better. Energy efficiency is generally the cheapest approach, and remaining energy needs can then be met with greener energy supply.

In addition to the set of principles shown above, municipal, national and state governments have different roles and degrees of influence in achieving ZCB pathways. These guiding principles lay the groundwork for a menu of pathways to arrive at net zero carbon for individual buildings, districts and building portfolios – turning carbon neutrality from an aspiration into a target well within reach.

Zero Carbon Buildings for a Sustainable Future

Fossil fuel-based energy consumption is still dominant in the world today, and there is a consensus on the limited reserves of these energy resources. Therefore, there is a strong stimulation into clean energy technologies to narrow the gap between fossil fuels and renewables. In this respect, several commitments and codes are proposed and adopted for a low energy-consuming world and for desirable environmental conditions. Sectoral energy consumption analyses clearly indicate that buildings are of vital importance in terms of energy consumption figures. From this point of view, buildings have a great potential for decisive and urgent reduction of energy consumption levels and thus greenhouse gas (GHG) emissions. Among the available retrofit solutions, greenery systems (GSs) stand for a reliable, cost-effective and eco-friendly method for remarkablemitigation of energy consumed in buildings. Through the works comparing the thermal regulation performance of uninsulated and green roofs, it is observed that the GS provides 20°C lower surface temperature in operation. Similar to green roofs, vertical greenery systems (VGSs) also reduce energy demand to approximately 25% as a consequence of wind blockage effects in winter.

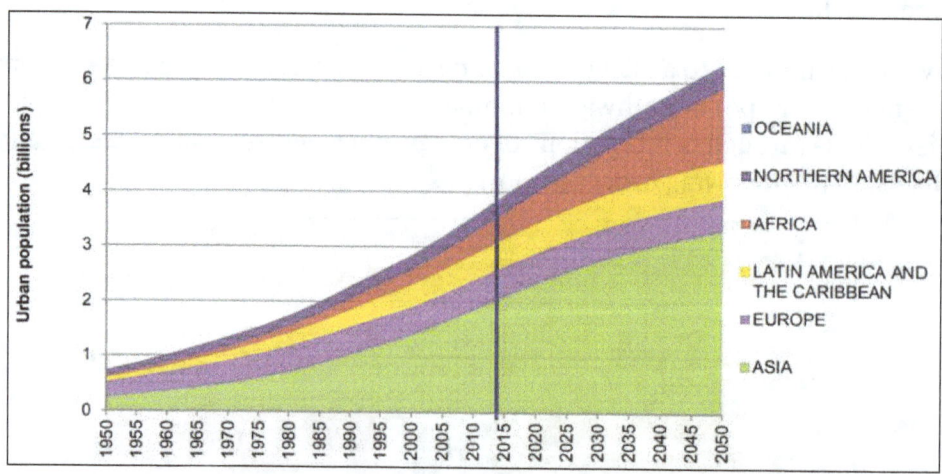

Urban population by major area, 1950–2050.

Since the beginning of the industrial revolution (roughly 200 years), the dramatic increase in world population and technological advancements led to remarkable rises in global energy demand. Scientists address a relationship between the global energy demand and the consumption of natural resources through the economic growth across the world, especially over the last two decades. Uncontrolled energy consumption due to human activities plays a vital role in biodiversity decline. The greatest part of the decline in biodiversity has taken place within the last 50 years. Urbanization is another significant problem of today's world in terms of growing importance of environmental issues. The urbanization rate is to rise by 75% until 2030 as shown in figure. Urbanization-related environmental matters can be illustrated as pollution, the depletion of natural resources, climate change, and global warming. Especially climate change notably affects the biotic systems as it has cumulative impacts on the global environment such as terrible weather conditions and deterioration of natural ecosystem (serious decrease in fishery stocks and in the productivity of lands).

The European Commission primarily aims to slow down the increase in greenhouse gas (GHG) emission to prevent the hazardous impacts on the environment. Based on the roadmap reported by European Commissions in 2010, the abatement in the EU GHG emissions is aimed to be 80% by 2050 (as compared to the 1990 level). The target of the decrease in GHG emissions takes place in the range 25–60% between 2020 and 2040. To reach this goal, the increase in the global temperature should be 2°C less than the pre-industrial era. A similar, national plan underlying the significance of climate change is adopted by the Government of China. Based on this plan, carbon emissions are expected to be reduced by 40–50% until 2020 compared to the level of 2005. However, it is a clear challenge to achieve the said targets concerned with global warming and GHG emissions. In this respect, appropriate investments in energy, transport, industry, information technologies, and building sectors are required for the desired outputs. Among the relevant sectors, buildings stand for the most promising field in terms of eco-friendly-mitigating energy consumption levels. The reduction of energy consumed in buildings does not have any negative effects on the welfare of the dwellers. L/ZCB strategy can be accomplished by constructing new environmentally friendly building or retrofitting existing buildings with low-cost, energy-efficient, and eco-friendly technologies. The retrofit of existing buildings can remarkably reduce energy demands and carbon emissions, as well as mitigating the depletion of natural resources. GSs are considered as low-energy concept for buildings, and they can be deployed in existing buildings as retrofit applications.

CO_2 Emissions

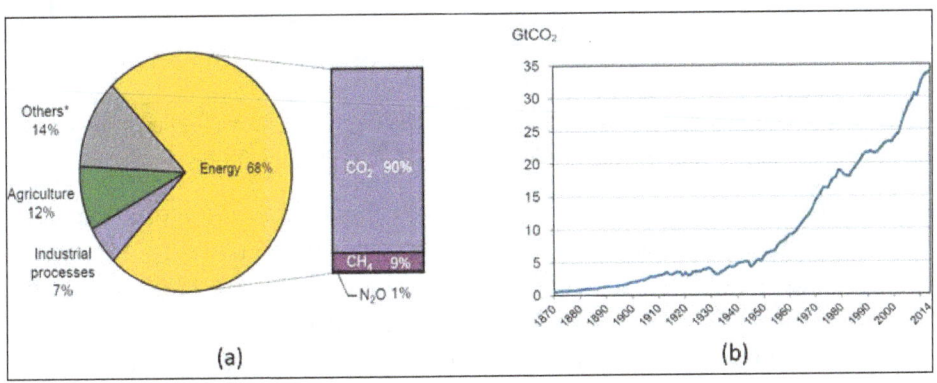

Estimated shares of global anthropogenic GHG, 2014 (a), trend in CO_2 emissions from fossil fuel combustion, 1870–2014 (b).

With respect to consensus among scientists, CO_2 emissions in the atmosphere have a remarkably rising trend since industrial revolution. In comparison to pre-industrial revolution, the average rise in CO_2 concentration with 403 ppm is reported to be about 40%. Depending on the recent assessment report on climate change, human beings have a considerable influence on the climate system due to the energy consumption. Therefore, the energy usage is admitted to be the greatest contributor to GHG emissions. Figure illustrates the shares of global GHG based on human activity.

The level of CO_2 emission is represented in figure. It is clear from the data that the CO2 emissions have a steadily rising trend from industrial revolution up to 2014. It is reported by Boeck et al. that the emissions are expected to increase to 52% from 2005 to 2050 if no decisive measures are taken. During the said period, carbon emissions are predicted to increase by 78%, which is notable. Also the annual increase in GHG emissions between 2000 and 2010 is found to be 1 giga tone of CO2 equivalent. When the emissions from 1970 to 2010 are analyzed, the growth is reported to be around 0.4 GtCO2eq. Moreover, carbon emissions dramatically increase with the explosive growth in global economy and world population. On the other hand, within the last decade, the emissions have a decreasing tendency because of the global economic recession between 2007 and 2008 as shown in figure. In 2015, global CO_2 emission level is predicted to be 32.3 $GtCO_2$, which is 0.1% lower than the level in 2014. For 2013 and 2014, the growth rate of CO_2 emission is given to be 1.7 and 0.6%, respectively. On the other hand, the annual rise of the emissions is reported to be 2.2% since 2000. From this point of view, it can be easily understood that the growth in global economy is independent of the reductions in GHG emissions.

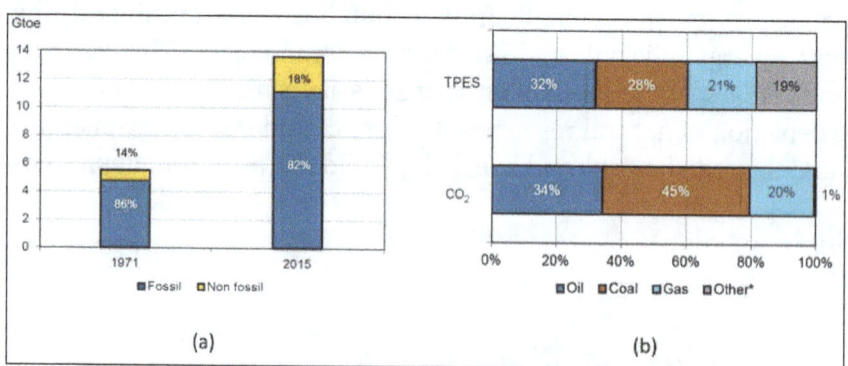

Energy supply by fossil and nonfossil fuels (a), world primary energy
supply and CO2 emissions: Shares by fuel in 2015 (b).

As a consequence of rising welfare of the countries at growing economic indicators, global energy demand remarkably increases. Between 1971 and 2015, the rise of global energy demand is reported to be 150%. While expressing the energy demand, total primary energy supply (TPES) is widely used to determine the rates as shown in figure. With respect to the emissions from fuel combustion in 2015, the largest share of CO_2 emissions is attributed to coal. However, the percentage of coal consumption (28%) is lower than the oil consumption (32%) according to the TPES data.

The major CO_2 emission sectors are electricity and heat generation, which are responsible for 42% of the total emissions in 2015. Although the share of oil utilized in electricity and heat generation decreases, the coal and gas consumptions have an increasing trend in 2015 as compared to the year of 1990 as depicted in figure.

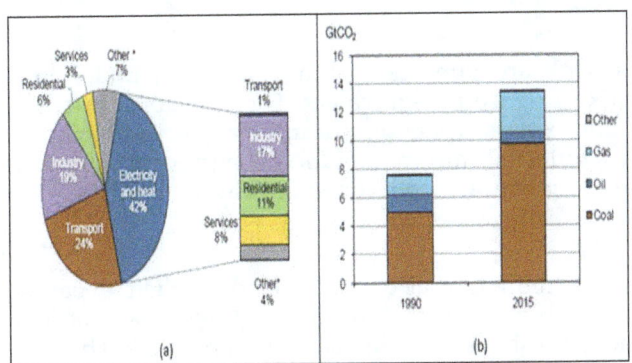

World CO_2 emissions from fuel combustion by sector, 2015 (a), CO_2 emissions from electricity and heat generation, 1990–2015 (b).

References

- "Raising The Bar - Building sustainable business value through environmental targets". Carbon Trust. June 2011. Retrieved 2012-11-12

- Reducing-co2-emission: thebalancesmb.com, Retrieved 16 May, 2020

- Joshi, S (2000). "Product Environmental Life Cycle Assessment Using Input-Output Techniques". Journal of Industrial Ecology. 3 (2–3): 95–120. Doi:10.1162/108819899569449

- Bowen, Frances; Wittneben, Bettina (2011). "Carbon accounting". Accounting, Auditing & Accountability Journal. 24 (8): 1022–1036. Doi:10.1108/09513571111184742

- Sweet, Cassandra (2010-01-06). "UPDATE: Global Carbon Trading Up In 2009, Though Prices Lower". The Wall Street Journal. Retrieved 2010-03-03

- Gillenwater, Michael; Derik Broekhoff; Mark Trexler; Jasmine Hyman; Rob Fowler (2007). "Policing the voluntary carbon market". Nature Reports Climate Change. 6 (711): 85–87. Doi:10.1038/climate.2007.58

- Sweet, Cassandra (2010-01-06). "UPDATE: Global Carbon Trading Up In 2009, Though Prices Lower". The Wall Street Journal. Retrieved 2010-03-03

INDEX